> SPSS 14.0 Brief Guide

For more information about SPSS® software products, please visit our Web site at *http://www.spss.com* or contact

SPSS Inc.
233 South Wacker Drive, 11th Floor
Chicago, IL 60606-6412
Tel: (312) 651-3000
Fax: (312) 651-3668

Preface

The *SPSS® 14.0 Brief Guide* provides a set of tutorials designed to acquaint you with the various components of the SPSS system. You can work through the tutorials in sequence or turn to the topics for which you need additional information. You can use this guide as a supplement to the online tutorial that is included with the SPSS Base 14.0 system or ignore the online tutorial and start with the tutorials found here.

SPSS 14.0

SPSS 14.0 is a comprehensive system for analyzing data. SPSS can take data from almost any type of file and use them to generate tabulated reports, charts, and plots of distributions and trends, descriptive statistics, and complex statistical analyses.

SPSS makes statistical analysis more accessible for the beginner and more convenient for the experienced user. Simple menus and dialog box selections make it possible to perform complex analyses without typing a single line of command syntax. The Data Editor offers a simple and efficient spreadsheet-like facility for entering data and browsing the working data file.

Internet Resources

The SPSS Web site (*http://www.spss.com*) offers answers to frequently asked questions about installing and running SPSS software and provides access to data files and other useful information.

In addition, the SPSS USENET discussion group (not sponsored by SPSS) is open to anyone interested in SPSS products. The USENET address is *comp.soft-sys.stat.spss*. It deals with computer, statistical, and other operational issues related to SPSS software.

You can also subscribe to an e-mail message list that is gatewayed to the USENET group. To subscribe, send an e-mail message to *listserv@listserv.uga.edu*. The text of the e-mail message should be: subscribe SPSSX-L firstname lastname. You can then post messages to the list by sending an e-mail message to *listserv@listserv.uga.edu*.

Additional Publications

For additional information about the features and operations of SPSS Base 14.0, you can consult the *SPSS Base 14.0 User's Guide*, which includes information on standard graphics. Examples using the statistical procedures found in SPSS Base 14.0 are provided in the Help system, installed with the software. Algorithms used in the statistical procedures are available on the product CD-ROM.

In addition, beneath the menus and dialog boxes, SPSS uses a command language. Some extended features of the system can be accessed only via command syntax. (Those features are not available in the Student Version.) Complete command syntax is documented in the *SPSS 14.0 Command Syntax Reference*, available in PDF form from the Help menu.

Individuals worldwide can order additional product manuals directly from SPSS Inc. For telephone orders in the United States and Canada, call SPSS Inc. at 800-543-2185. For telephone orders outside of North America, contact your local office, listed on the SPSS Web site at *http://www.spss.com/worldwide*.

The *SPSS Statistical Procedures Companion*, by Marija Norušis, has been published by Prentice Hall. It contains overviews of the procedures in the SPSS Base, plus Logistic Regression and General Linear Models. The *SPSS Advanced Statistical Procedures Companion* has also been published by Prentice Hall. It includes overviews of the procedures in the SPSS Advanced and Regression modules.

SPSS Options

The following options are available as add-on enhancements to the full (not Student Version) SPSS Base system:

SPSS Regression Models™ provides techniques for analyzing data that do not fit traditional linear statistical models. It includes procedures for probit analysis, logistic regression, weight estimation, two-stage least-squares regression, and general nonlinear regression.

SPSS Advanced Models™ focuses on techniques often used in sophisticated experimental and biomedical research. It includes procedures for general linear models (GLM), linear mixed models, variance components analysis, loglinear analysis, ordinal regression, actuarial life tables, Kaplan-Meier survival analysis, and basic and extended Cox regression.

SPSS Tables™ creates a variety of presentation-quality tabular reports, including complex stub-and-banner tables and displays of multiple response data.

SPSS Trends™ performs comprehensive forecasting and time series analyses with multiple curve-fitting models, smoothing models, and methods for estimating autoregressive functions.

SPSS Categories® performs optimal scaling procedures, including correspondence analysis.

SPSS Conjoint™ performs conjoint analysis.

SPSS Exact Tests™ calculates exact p values for statistical tests when small or very unevenly distributed samples could make the usual tests inaccurate.

SPSS Missing Value Analysis™ describes patterns of missing data, estimates means and other statistics, and imputes values for missing observations.

SPSS Maps™ turns your geographically distributed data into high-quality maps with symbols, colors, bar charts, pie charts, and combinations of themes to present not only what is happening but where it is happening.

SPSS Complex Samples™ allows survey, market, health, and public opinion researchers, as well as social scientists who use sample survey methodology, to incorporate their complex sample designs into data analysis.

SPSS Classification Tree™ creates a tree-based classification model. It classifies cases into groups or predicts values of a dependent (target) variable based on values of independent (predictor) variables. The procedure provides validation tools for exploratory and confirmatory classification analysis.

SPSS Data Validation™ provides a quick visual snapshot of your data. It provides the ability to apply validation rules that identify invalid data values. You can create rules that flag out-of-range values, missing values, or blank values. You can also save variables that record individual rule violations and the total number of rule violations per case. A limited set of predefined rules that you can copy or modify is provided.

The SPSS family of products also includes applications for data entry, text analysis, classification, neural networks, and flowcharting.

Training Seminars

SPSS Inc. provides both public and onsite training seminars for SPSS. All seminars feature hands-on workshops. SPSS seminars will be offered in major U.S. and European cities on a regular basis. For more information on these seminars, contact your local office, listed on the SPSS Web site at *http://www.spss.com/worldwide*.

Technical Support

The services of SPSS Technical Support are available to registered customers of SPSS. (Student Version customers should read the special section on technical support for the Student Version. For more information, see "Technical Support for Students" on p. vii.) Customers may contact Technical Support for assistance in using SPSS products or for installation help for one of the supported hardware environments. To reach Technical Support, see the SPSS Web site at *http://www.spss.com*, or contact your local office, listed on the SPSS Web site at *http://www.spss.com/worldwide*. Be prepared to identify yourself, your organization, and the serial number of your system.

Tell Us Your Thoughts

Your comments are important. Please let us know about your experiences with SPSS products. We especially like to hear about new and interesting applications using the SPSS system. Please send e-mail to *suggest@spss.com*, or write to SPSS Inc., Attn: Director of Product Planning, 233 South Wacker Drive, 11th Floor, Chicago IL 60606-6412.

SPSS 14.0 for Windows Student Version

The SPSS 14.0 for Windows Student Version is a limited but still powerful version of the SPSS Base 14.0 system.

Capability

The Student Version contains all of the important data analysis tools contained in the full SPSS Base system, including:

- Spreadsheet-like Data Editor for entering, modifying, and viewing data files.
- Statistical procedures, including *t* tests, analysis of variance, and crosstabulations.
- Interactive graphics that allow you to change or add chart elements and variables dynamically; the changes appear as soon as they are specified.
- Standard high-resolution graphics for an extensive array of analytical and presentation charts and tables.

Limitations

Created for classroom instruction, the Student Version is limited to use by students and instructors for educational purposes only. The Student Version does not contain all of the functions of the SPSS Base 14.0 system. The following limitations apply to the SPSS 14.0 for Windows Student Version:

- Data files cannot contain more than 50 variables.

- Data files cannot contain more than 1,500 cases. SPSS add-on modules (such as Regression Models or Advanced Models) cannot be used with the Student Version.

- SPSS command syntax is not available to the user. This means that it is not possible to repeat an analysis by saving a series of commands in a syntax or "job" file, as can be done in the full version of SPSS.

- Scripting and automation are not available to the user. This means that you cannot create scripts that automate tasks that you repeat often, as can be done in the full version of SPSS.

Technical Support for Students

Students should obtain technical support from their instructors or from local support staff identified by their instructors. Technical support from SPSS for the SPSS 14.0 Student Version is provided *only to instructors using the system for classroom instruction.*

Before seeking assistance from your instructor, please write down the information described below. Without this information, your instructor may be unable to assist you:

- The type of PC you are using, as well as the amount of RAM and free disk space you have.

- The operating system of your PC.

- A clear description of what happened and what you were doing when the problem occurred. If possible, please try to reproduce the problem with one of the sample data files provided with the program.

- The exact wording of any error or warning messages that appeared on your screen.

- How you tried to solve the problem on your own.

Technical Support for Instructors

Instructors using the Student Version for classroom instruction may contact SPSS Technical Support for assistance. In the United States and Canada, call SPSS Technical Support at 312-651-3410, or send an e-mail to *support@spss.com*. Please include your name, title, and academic institution.

Instructors outside of the United States and Canada should contact your local SPSS office, listed on the SPSS Web site at *http://www.spss.com/worldwide*.

Contents

4   Using the Data Editor   55

5   Working with Multiple Data Sources   81

8 Creating and Editing Charts 135

9 Working with Syntax 187

10 Modifying Data Values
193

11 Sorting and Selecting Data
215

12 Additional Statistical Procedures
227

Introduction

This guide provides a set of tutorials designed to acquaint you with the various components of the SPSS system. You can work through the tutorials in sequence or turn to the topics for which you need additional information. The goal is to enable you to perform useful analyses on your data using SPSS.

This chapter will introduce you to the basic environment of SPSS and demonstrate a typical session. We will run SPSS, retrieve a previously defined SPSS data file, and then produce a simple statistical summary and a chart. In the process, you will learn the roles of the primary windows within SPSS and will see a few features that smooth the way when you are running analyses.

More detailed instruction about many of the topics touched upon in this chapter will follow in later chapters. Here, we hope to give you a basic framework for understanding and using SPSS.

Sample Files

Most of the examples that are presented here use the data file *demo.sav*. This data file is a fictitious survey of several thousand people, containing basic demographic and consumer information.

All sample files that are used in these examples are located in the folder in which SPSS is installed or in the *tutorial\sample_files* folder within the SPSS installation folder.

Starting SPSS

To start SPSS:

▶ From the Windows Start menu choose:

Programs
  SPSS for Windows
    SPSS for Windows

To start SPSS for Windows Student Version:

▶ From the Windows Start menu choose:

Programs
  SPSS for Windows
    Student Version

When you start a session, you see the Data Editor window.

Figure 1-1
Data Editor window (Data View)

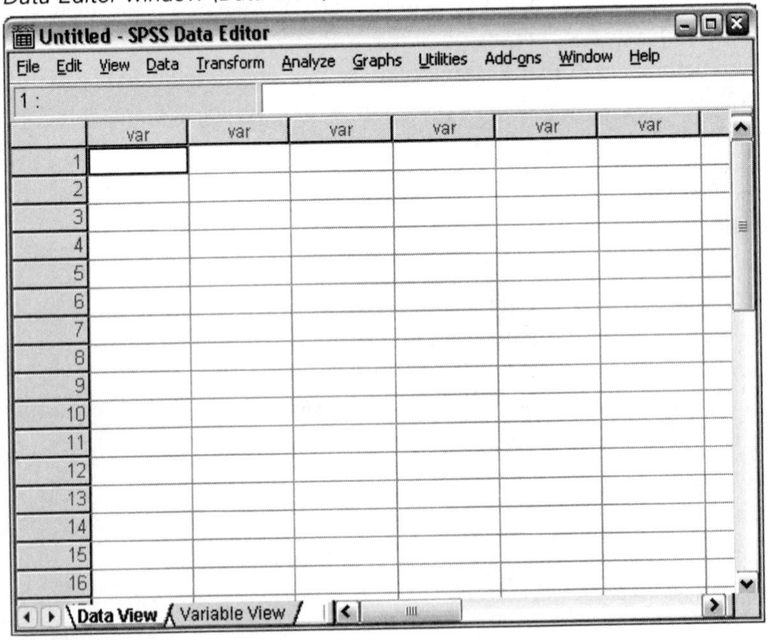

Variable Display in Dialog Boxes

Either variable names or longer variable labels will appear in list boxes in dialog boxes. Variables in list boxes can be ordered alphabetically or by their position in the file.

In this guide, we will display variable labels in alphabetical order within list boxes. For a new user of SPSS, this setup provides a more complete description of variables in an easy-to-follow order.

The default setting within SPSS is to display variable labels in file order. To change the order of variable labels before accessing data:

▶ From the menus choose:
  Edit
    Options...

▶ On the General tab, select Display labels in the Variable Lists group.

▶ Select Alphabetical.

▶ Click OK, and then click OK to confirm the change.

Opening a Data File

To open a data file:

▶ From the menus choose:
  File
    Open
      Data...

Alternatively, you can use the Open File button on the toolbar.

Figure 1-2
Open File toolbar button

The Open File dialog box is displayed.

Figure 1-3
Open File dialog box

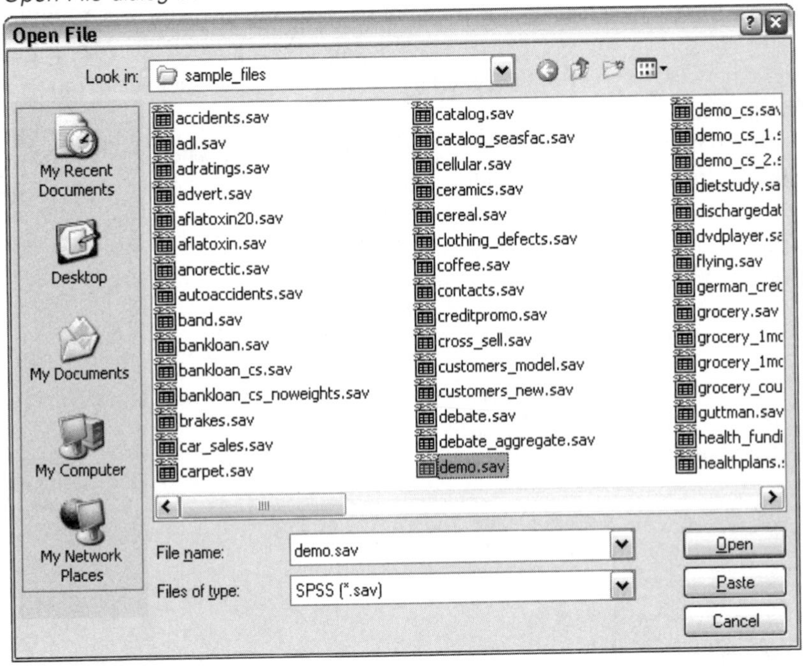

By default, SPSS-format data files (*.sav* extension) are displayed. You can display other file formats by using the Files of type drop-down list.

By default, data files in the folder (directory) in which SPSS is installed are displayed. The files for this guide are located in the folder in which SPSS is installed or in the *tutorial\sample_files* folder within the SPSS installation folder.

▶ Double-click the *tutorial* folder.

▶ Double-click the *sample_files* folder.

▶ Click the file *demo.sav* (or just *demo* if file extensions are not displayed).

▶ Click Open to open the SPSS data file.

Figure 1-4
demo.sav file in Data Editor

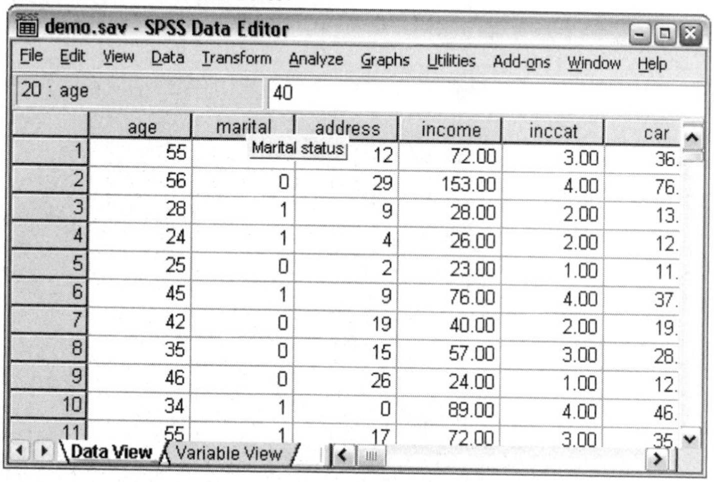

	age	marital	address	income	inccat	car
1	55	Marital status	12	72.00	3.00	36.
2	56	0	29	153.00	4.00	76.
3	28	1	9	28.00	2.00	13.
4	24	1	4	26.00	2.00	12.
5	25	0	2	23.00	1.00	11.
6	45	1	9	76.00	4.00	37.
7	42	0	19	40.00	2.00	19.
8	35	0	15	57.00	3.00	28.
9	46	0	26	24.00	1.00	12.
10	34	1	0	89.00	4.00	46.
11	55	1	17	72.00	3.00	35.

demo.sav - SPSS Data Editor

File Edit View Data Transform Analyze Graphs Utilities Add-ons Window Help

20 : age   40

Data View   Variable View

The data file is displayed in the Data Editor. The SPSS Viewer is also displayed, showing the dataset name. You can minimize the SPSS Viewer to display the Data Editor. In the Data Editor, if you put the mouse cursor on a variable name (the column headings), a more descriptive variable label is displayed (if a label has been defined for that variable).

By default, the actual data values are displayed. To display labels:

► From the menus choose:
View
    Value Labels

Alternatively, you can use the Value Labels button on the toolbar.

Figure 1-5
Value Labels button

Descriptive value labels are now displayed to make it easier to interpret the responses.

Figure 1-6

Value labels displayed in the Data Editor

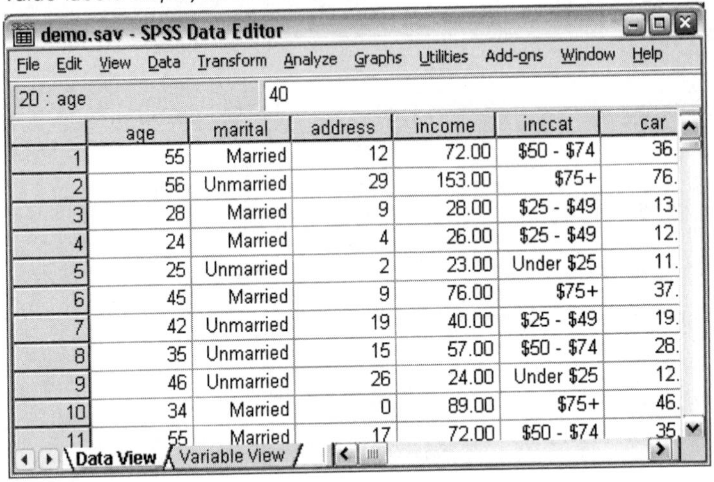

		age	marital	address	income	inccat	car
	1	55	Married	12	72.00	$50 - $74	36.
	2	56	Unmarried	29	153.00	$75+	76.
	3	28	Married	9	28.00	$25 - $49	13.
	4	24	Married	4	26.00	$25 - $49	12.
	5	25	Unmarried	2	23.00	Under $25	11.
	6	45	Married	9	76.00	$75+	37.
	7	42	Unmarried	19	40.00	$25 - $49	19.
	8	35	Unmarried	15	57.00	$50 - $74	28.
	9	46	Unmarried	26	24.00	Under $25	12.
	10	34	Married	0	89.00	$75+	46.
	11	55	Married	17	72.00	$50 - $74	35

Running an Analysis

The Analyze menu contains a list of general reporting and statistical analysis categories. Most of the categories are followed by an arrow, which indicates that there are several analysis procedures available within the category; these procedures will appear on a submenu when the category is selected.

We will start by creating a simple frequency table (table of counts).

► From the menus choose:

Analyze
  Descriptive Statistics
    Frequencies...

The Frequencies dialog box is displayed.

Figure 1-7
Frequencies dialog box

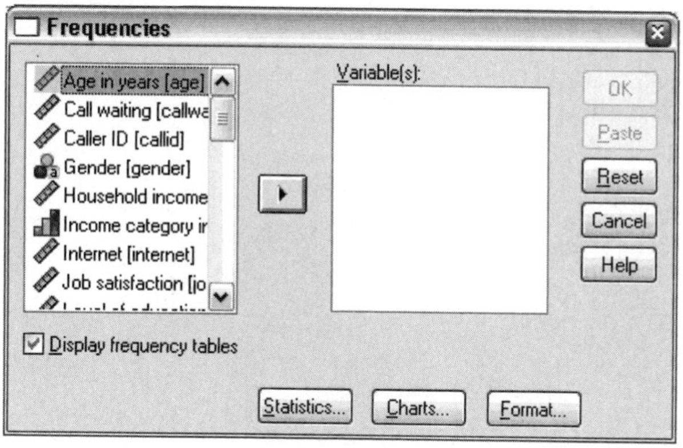

An icon next to each variable provides information about data type and level of measurement.

Measurement Level	Data Type			
	Numeric	String	Date	Time
Scale		n/a		
Ordinal				
Nominal				

▶ Click the variable *Income category in thousands [inccat]*.

Figure 1-8
Variable labels and names in the Frequencies dialog box

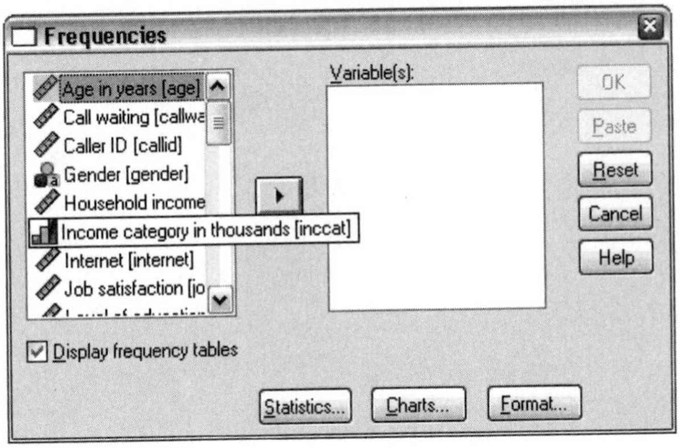

A more complete description of each variable pops up when the cursor is over it. The variable name for *Income category in thousands* (in square brackets) is *inccat*, and it has the variable label *Income category in thousands [inccat]*. If there were no variable label, only the variable name would appear in the list box.

In the dialog box, you choose the variables that you want to analyze from the source list on the left and move them into the Variable(s) list on the right. The OK button, which runs the analysis, is disabled until at least one variable is placed in the Variable(s) list.

You can obtain additional labeling information by right-clicking any variable name in the list.

▶ Right-click *Income category in thousands [inccat]* and choose Variable Information.

▶ Click the down arrow on the Value labels drop-down list.

Figure 1-9
Defined labels for income variable

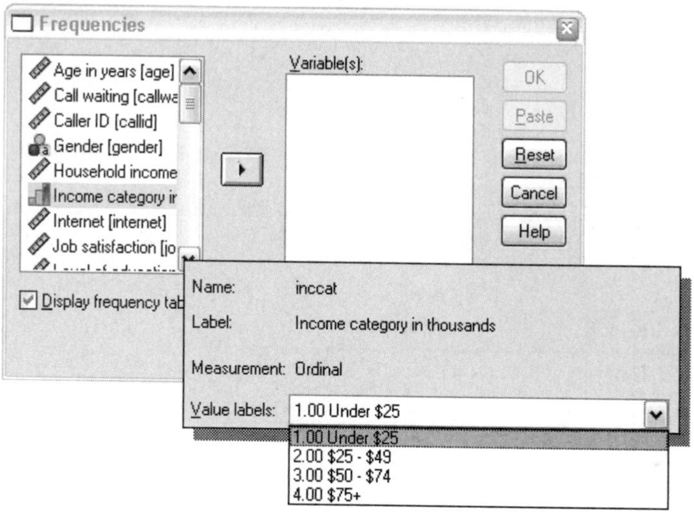

All of the defined value labels for the variable are displayed.

▶ Click *Gender [gender]* in the source variable list, and then click the right-arrow button to move the variable into the target Variable(s) list.

▶ Click *Income category in thousands [inccat]* in the source list, and then click the right-arrow button again.

Figure 1-10
Variables selected for analysis

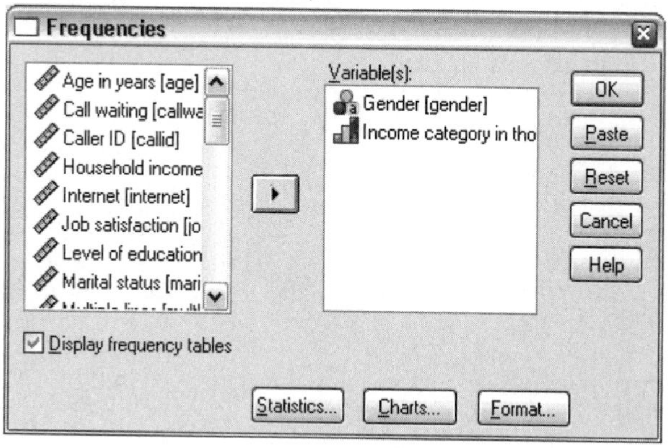

▶ Click OK to run the procedure.

Viewing Results

Figure 1-11
Viewer window

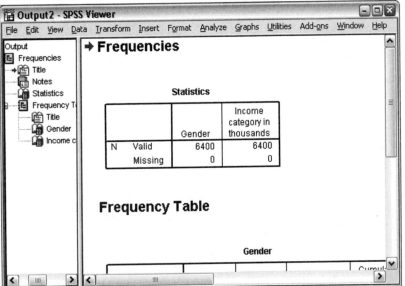

Results are displayed in the Viewer window.

You can quickly go to any item in the Viewer by selecting it in the outline pane.

▶ Click Income category in thousands [inccat].

Figure 1-12
Frequency table of income categories

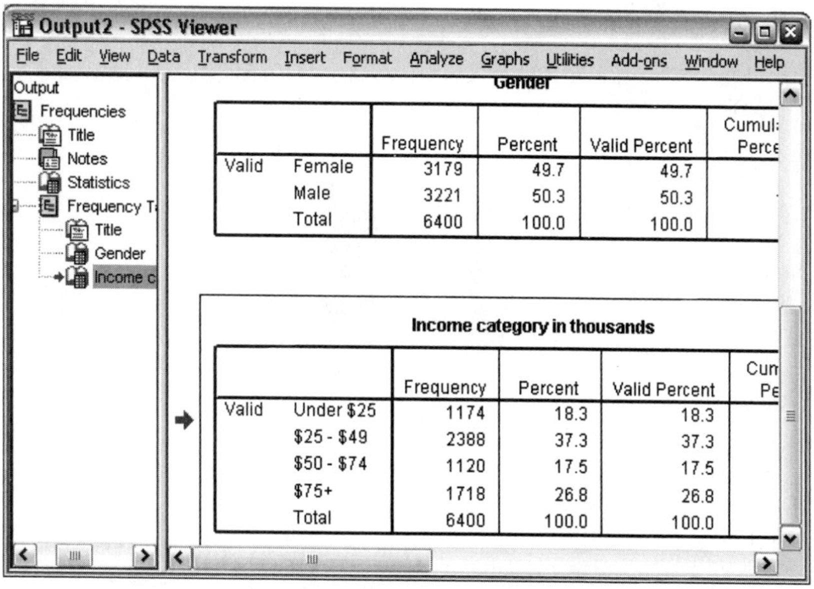

The frequency table for income categories is displayed. This frequency table shows the number and percentage of people in each income category.

Creating Charts

Although some statistical procedures can create high-resolution charts, you can also use the Graphs menu to create charts.

For example, you can create a chart that shows the relationship between wireless telephone service and PDA (personal digital assistant) ownership.

► From the menus choose:
Graphs
  Chart Builder...

► Click the Gallery tab (if it is not selected).

► Click Bar (if it is not selected).

► Drag the Clustered Bar icon onto the canvas, which is the large area above the Gallery.

Figure 1-13
Chart Builder dialog box

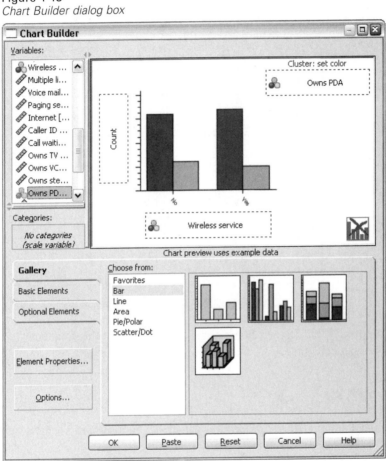

▶ Scroll down the Variables list, right-click *Wireless service [wireless]*, and then choose Nominal as its measurement level.

▶ Drag the *Wireless service [wireless]* variable to the *x* axis.

▶ Right-click *Owns PDA [ownpda]* and choose Nominal as its measurement level.

▶ Drag the *Owns PDA [ownpda]* variable to the cluster drop zone in the upper right corner of the canvas.

▶ Click OK to create the chart.

Figure 1-14
Bar chart displayed in Viewer window

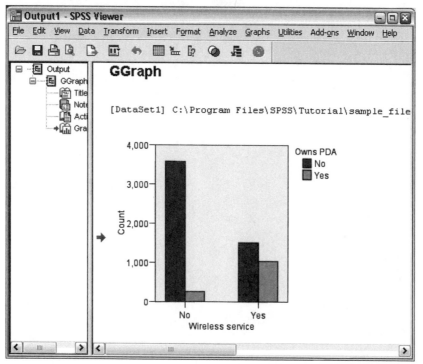

The bar chart is displayed in the Viewer. The chart shows that people with wireless phone service are far more likely to have PDAs than people without wireless service.

You can edit charts and tables by double-clicking them in the contents pane of the Viewer window, and you can copy and paste your results into other applications. Those topics will be covered later.

Exiting SPSS

To exit SPSS:

▶ From the menus choose:
File
  Exit

▶ Click No if you get an alert asking if you want to save your results.

Using the Help System

Help is available in a number of different ways, including:

Help menu. Every window has a Help menu on the menu bar. The Topics menu item provides access to the Help system, where you can use the Contents and Index tabs to find topics. The Tutorial menu item provides access to the introductory tutorial.

Dialog box Help buttons. Most dialog boxes have a Help button that takes you directly to a Help topic for that dialog box. The Help topic provides general information and links to related topics.

Pivot table context menu Help. Right-click on terms in an activated pivot table in the Viewer and select What's This? from the context menu to display definitions of the terms.

Statistics Coach. The Statistics Coach item on the Help menu provides a wizard-like method for finding the right statistical or charting procedure for what you want to do.

Case Studies. The Case Studies item on the Help menu provides hands-on examples of how to create various types of statistical analyses and interpret the results. The sample data files used in the examples are also provided so that you can work through the examples to see exactly how the results were produced.

Microsoft Internet Explorer Settings

Most Help features in this application use technology based on Microsoft Internet Explorer. Some versions of Internet Explorer (including the version provided with Microsoft XP, Service Pack 2) will by default block content from what it interprets to be "pop-up" windows. This default setting may result in some blocked content in Help features. To see all Help content, you can change the default behavior of Internet Explorer.

▶ From the Internet Explorer menus choose:
Tools
  Internet Options...

▶ Click the Advanced tab.

▶ Scroll down to the Security section.

▶ Select (check) Allow active content to run in files on My Computer.

Files

This chapter uses the files *demo.sav* and *bhelptut.spo*.

Help Contents Tab

The Topics item on the Help menu opens a Help window.

▶ From the menus choose:
Help
  Topics

Figure 2-1
Help Contents tab

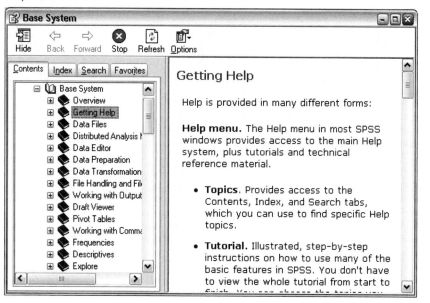

The Contents tab in the left pane of the Help window is an expandable and collapsible table of contents. It is most useful if you're looking for general information or are unsure of what index term to use to find what you're looking for.

Help Index Tab

Figure 2-2
Help Index tab

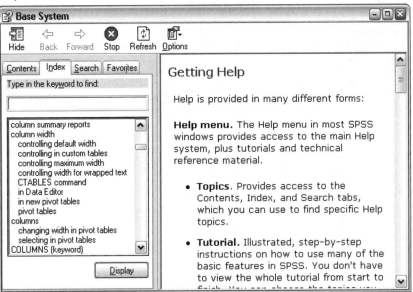

► Click the Index tab in the left pane of the Help window.

The Index tab provides a searchable index that makes it easy to find specific topics. The Index tab is organized in alphabetical order, just like a book index. It uses **incremental search** to find what you're looking for.

For example, you can:

▶ Type med.

Figure 2-3
Incremental index search

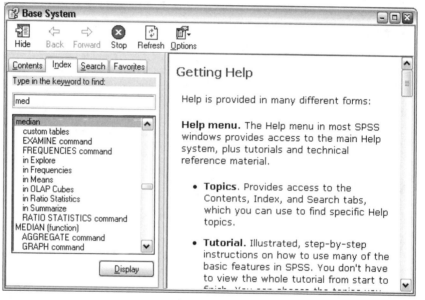

The index scrolls to and highlights the first index entry that starts with these letters, which is median.

Dialog Box Help

Most dialog boxes have a Help button that displays a Help topic about what the dialog box does and how to do it.

▶ From the menus choose:
Analyze
  Descriptive Statistics
    Frequencies...

▶ Click Help.

Figure 2-4
Dialog box Help topic

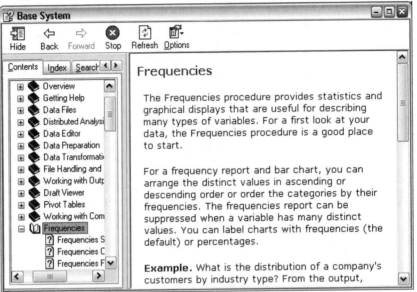

In this example, the Help topic describes the purpose of the Frequencies procedure and provides an example.

Statistics Coach

The Statistics Coach can help to guide you through the process of finding the procedure that you want to use.

▶ From the menus choose:
Help
  Statistics Coach

Figure 2-5
Statistics Coach, first step

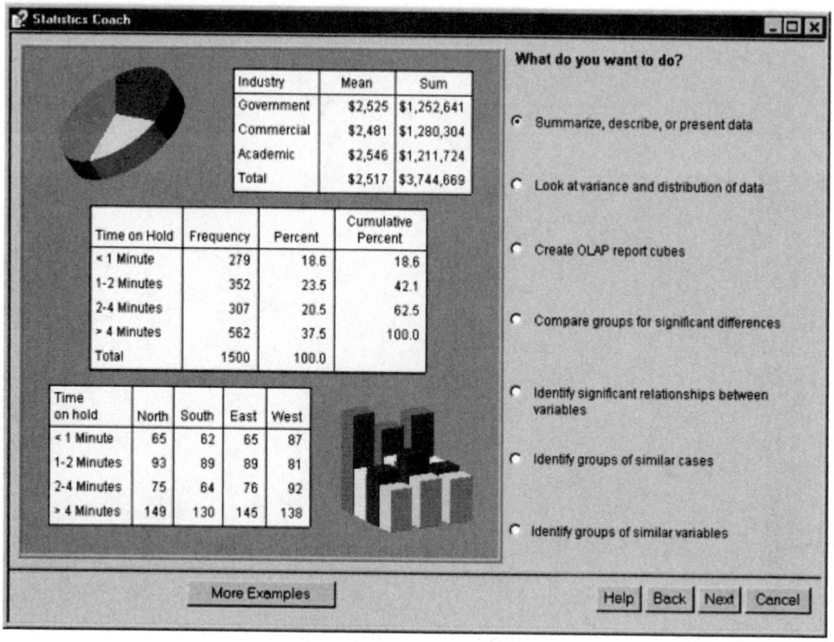

The Statistics Coach presents a series of questions designed to find the appropriate procedure. The first question is simply "What do you want to do?"

For example, if you want to summarize data:

▶ Select Summarize, describe, or present data.

▶ Then click Next.

Figure 2-6
Selecting a data type

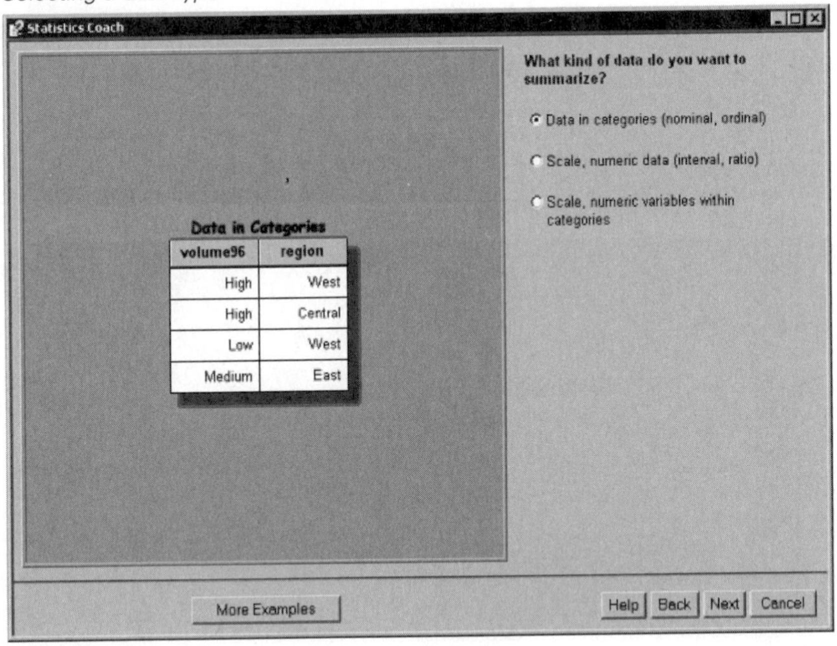

The next question asks about the type of data you want to summarize. If you're unsure, each choice displays different examples.

▶ Select Scale, numeric data (interval, ratio).

Figure 2-7
Selecting a different data type

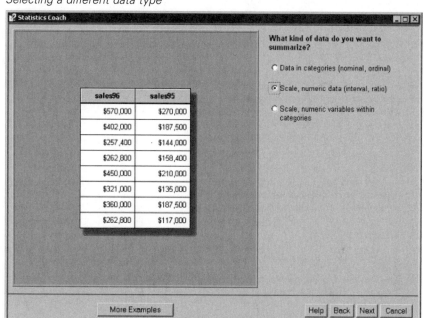

The example changes to reflect your choice. If you're still unsure, you can:

▶ Click More Examples.

A new example of the same data type is displayed. If the examples don't provide enough information, you can:

▶ Click Help.

Figure 2-8
Statistics Coach Help topic

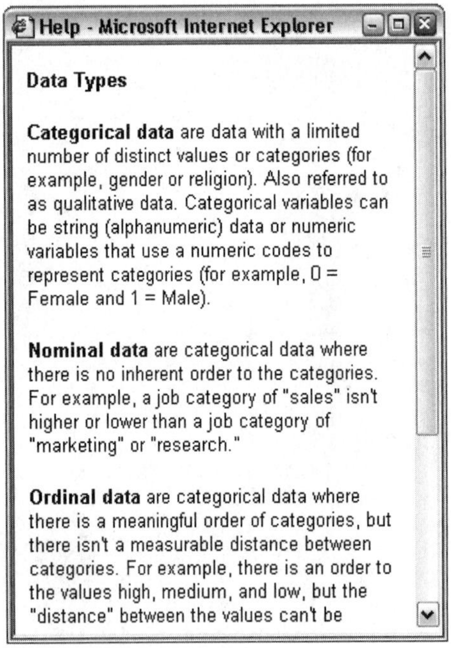

In this example, the Help topic defines the different data types.

▶ Close the Help window.

▶ Select Data in categories (nominal, ordinal) and click Next.

The next question asks you how you want to display your data.

▶ Select Tables and numbers and click Next.

Figure 2-9
Selecting tables or charts

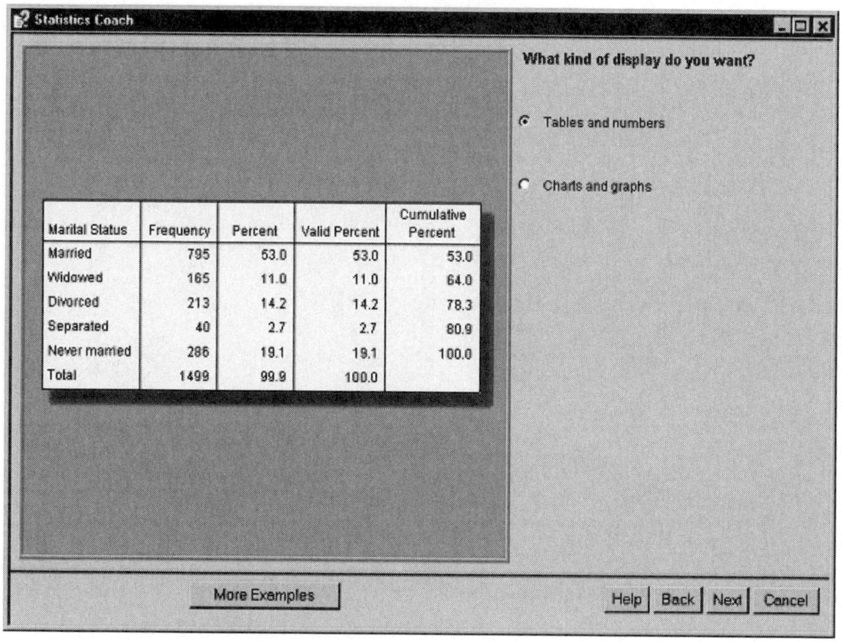
Statistics Coach
What kind of display do you want?
Tables and numbers
Charts and graphs

Marital Status	Frequency	Percent	Valid Percent	Cumulative Percent
Married	795	53.0	53.0	53.0
Widowed	165	11.0	11.0	64.0
Divorced	213	14.2	14.2	78.3
Separated	40	2.7	2.7	80.9
Never married	286	19.1	19.1	100.0
Total	1499	99.9	100.0	

More Examples
Help
Back
Next
Cancel

The final question asks you what kind of summary measure you want to display.

▶ Select Individual case listings within categories.

Figure 2-10
Statistics Coach, final step

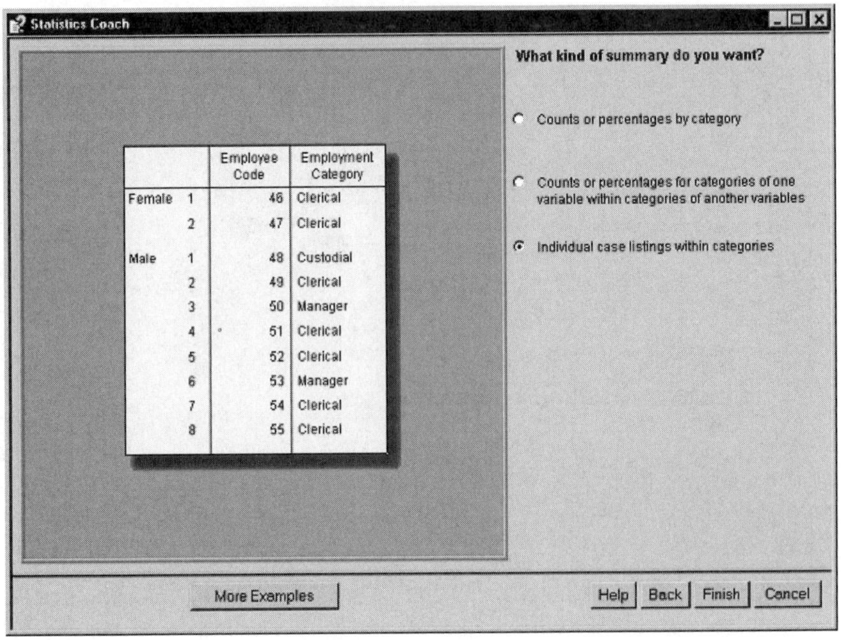

When the Statistics Coach has enough information, the Next button changes to Finish.

When you click Finish, the dialog box for the selected procedure opens automatically, and a Help topic for the procedure is also displayed.

Figure 2-11
Dialog box and Help topic

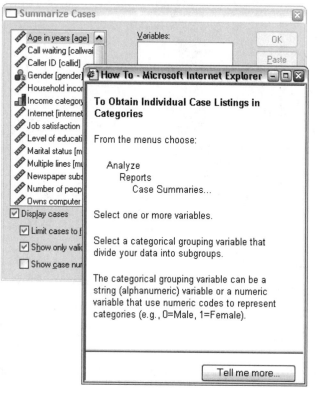

This is a custom Help topic, based on your selections in the Statistics Coach. Since some dialog boxes perform numerous functions, more than one path in the Statistics Coach may lead to the same dialog box, but the instructions in the Help topic may be different.

▶ Click Tell me more in the Help topic to get more detailed information.

Figure 2-12
"Tell me more" Help topic

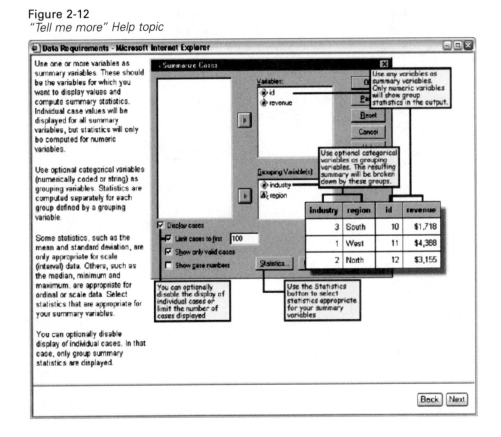

This Help topic provides detailed information on the data type(s) appropriate for the selected procedure.

Case Studies

Case studies provide comprehensive overviews of each procedure. Data files used in the examples are installed with SPSS, so you can follow along, performing the same analysis—from opening the data source and selecting variables for analysis to interpreting the results.

To access the case studies:

▶ Right-click on any pivot table created by a procedure. For example, you can right-click on the frequency table for Gender.

Figure 2-13
Accessing the case studies

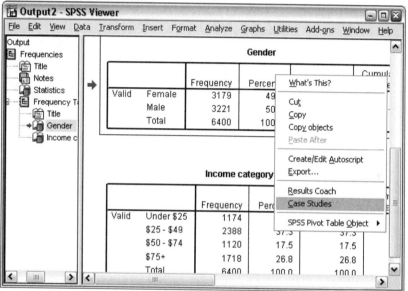

▶ Select Case Studies on the pop-up context menu.

Case studies are not available for all procedures. The Case Studies choice on the context menu will appear only if the feature is available for the procedure that created the selected pivot table.

Reading Data

Data can be entered directly into SPSS, or it can be imported from a number of different sources. The processes for reading data stored in SPSS data files, spreadsheet applications, such as Microsoft Excel, database applications, such as Microsoft Access, and text files are all discussed in this chapter.

Basic Structure of an SPSS Data File

Figure 3-1
Data Editor

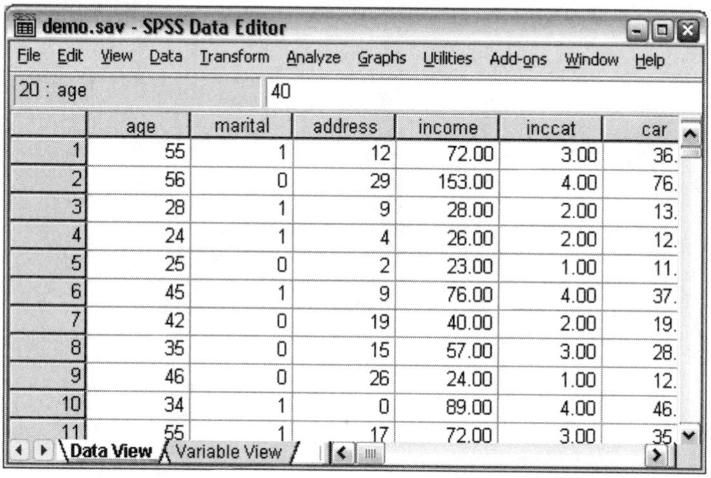

	age	marital	address	income	inccat	car
1	55	1	12	72.00	3.00	36.
2	56	0	29	153.00	4.00	76.
3	28	1	9	28.00	2.00	13.
4	24	1	4	26.00	2.00	12.
5	25	0	2	23.00	1.00	11.
6	45	1	9	76.00	4.00	37.
7	42	0	19	40.00	2.00	19.
8	35	0	15	57.00	3.00	28.
9	46	0	26	24.00	1.00	12.
10	34	1	0	89.00	4.00	46.
11	55	1	17	72.00	3.00	35.

SPSS data files are organized by cases (rows) and variables (columns). In this data file, cases represent individual respondents to a survey. Variables represent each question asked in the survey.

Reading an SPSS Data File

SPSS data files, which have a *.sav* file extension, contain your saved data. To open *demo.sav*, an example file that is installed with the product:

▶ From the menus choose:
File
  Open
    Data...

▶ Make sure that SPSS (*.sav) is selected in the Files of Type drop-down list.

Figure 3-2
Open File dialog box

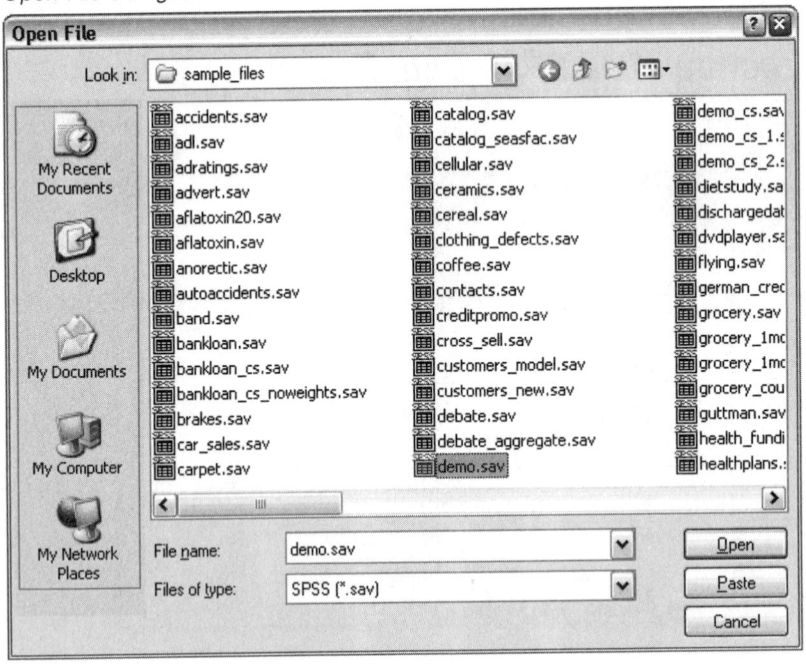

▶ Go to the *tutorial/sample_files* folder.

▶ Select *demo.sav* and click Open.

The data are now displayed in the Data Editor.

Figure 3-3
Opened data file

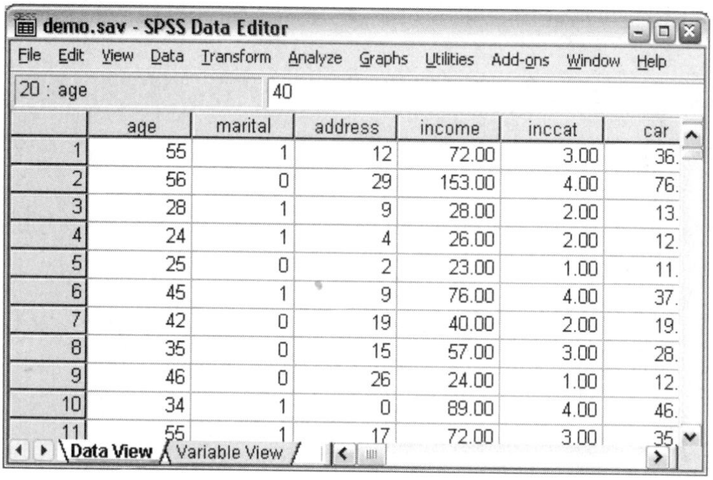

	age	marital	address	income	inccat	car
1	55	1	12	72.00	3.00	36.
2	56	0	29	153.00	4.00	76.
3	28	1	9	28.00	2.00	13.
4	24	1	4	26.00	2.00	12.
5	25	0	2	23.00	1.00	11.
6	45	1	9	76.00	4.00	37.
7	42	0	19	40.00	2.00	19.
8	35	0	15	57.00	3.00	28.
9	46	0	26	24.00	1.00	12.
10	34	1	0	89.00	4.00	46.
11	55	1	17	72.00	3.00	35.

Reading Data from Spreadsheets

Rather than typing all of your data directly into the Data Editor, you can read data from applications such as Microsoft Excel. You can also read column headings as variable names.

▶ From the menus choose:

File
  Open
    Data...

▶ Select Excel (*.xls) from the Files of Type drop-down list.

Figure 3-4
Open File dialog box

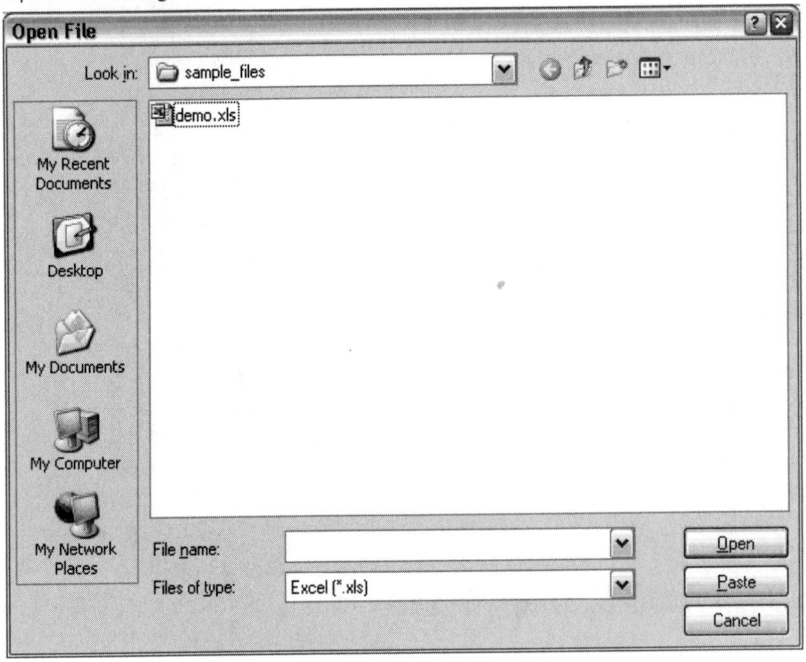

▶ Select *demo.xls* and click Open to read this spreadsheet.

The Opening Excel Data Source dialog box is displayed, allowing you to specify whether variable names are to be included in the spreadsheet, as well as the cells that you want to import. In Excel 5 or later, you can also specify which worksheets you want to import.

Figure 3-5
Opening Excel Data Source dialog box

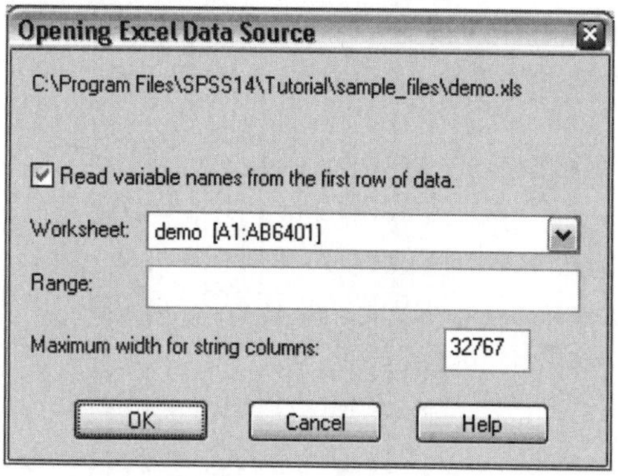

Opening Excel Data Source

C:\Program Files\SPSS14\Tutorial\sample_files\demo.xls

☑ Read variable names from the first row of data.

Worksheet: demo [A1:AB6401]

Range:

Maximum width for string columns: 32767

OK   Cancel   Help

▶ Make sure that Read variable names from the first row of data is selected. This option reads column headings as variable names.

If the column headings do not conform to the SPSS variable-naming rules, they are converted into valid variable names and the original column headings are saved as variable labels. If you want to import only a portion of the spreadsheet, specify the range of cells to be imported in the Range text box.

▶ Click OK to read the Excel file.

The data now appear in the Data Editor, with the column headings used as variable names. Since variable names can't contain spaces, the space from the original column headings have been removed. For example, *Marital status* in the Excel file becomes the variable *Maritalstatus* in SPSS. The original column heading is retained as a variable label.

Figure 3-6
Imported Excel data

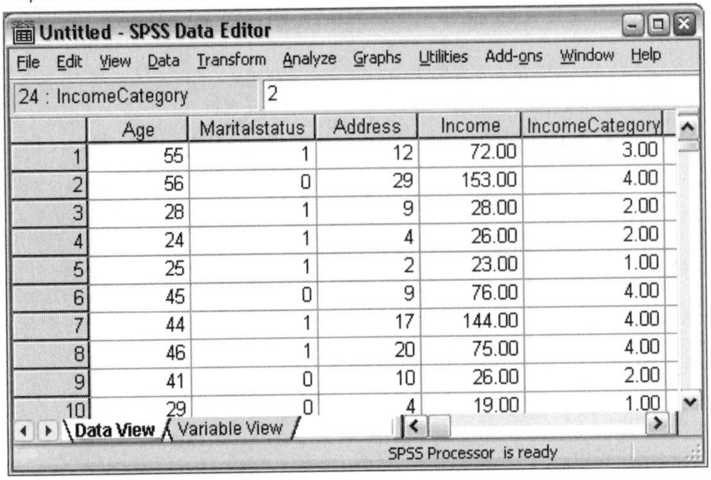

	Age	Maritalstatus	Address	Income	IncomeCategory
1	55	1	12	72.00	3.00
2	56	0	29	153.00	4.00
3	28	1	9	28.00	2.00
4	24	1	4	26.00	2.00
5	25	1	2	23.00	1.00
6	45	0	9	76.00	4.00
7	44	1	17	144.00	4.00
8	46	1	20	75.00	4.00
9	41	0	10	26.00	2.00
10	29	0	4	19.00	1.00

Reading Data from a Database

Data from database sources are easily imported using the Database Wizard. Any database that uses ODBC (Open Database Connectivity) drivers can be read directly by SPSS after the drivers are installed. ODBC drivers for many database formats are supplied on the installation CD. Additional drivers can be obtained from third-party vendors. One of the most common database applications, Microsoft Access, is discussed in this example.

▶ From the menus choose:

File
  Open Database
    New Query...

Figure 3-7
Database Wizard Welcome dialog box

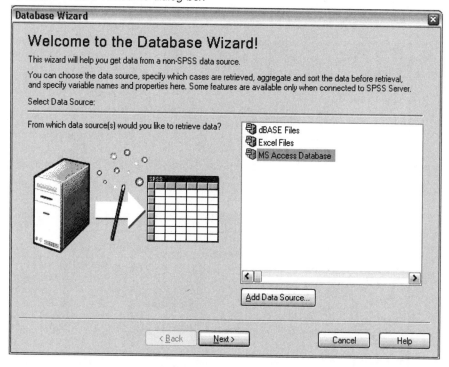

▶ Select MS Access Database from the list of data sources and click Next.

If MS Access Database is not listed here, you need to run *Microsoft Data Access Pack.exe*, which can be found in the Microsoft Data Access Pack folder on the CD.

Note: Depending on your installation, you may also see a list of OLEDB data sources on the left side of the wizard, but this example uses the list of ODBC data source displayed on the right side.

Figure 3-8
ODBC Driver Login dialog box

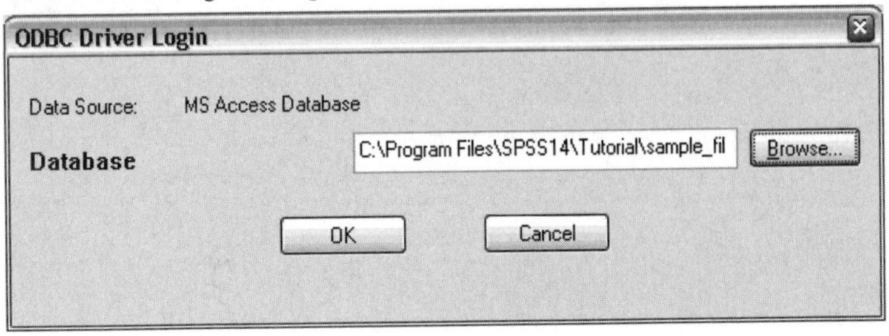

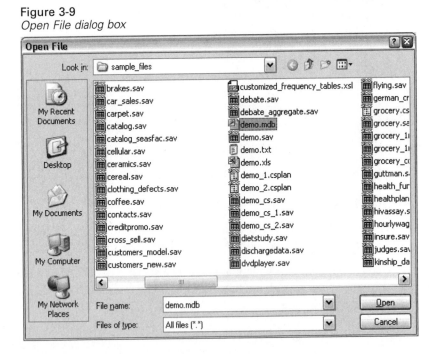

▶ Click Browse to navigate to the Access database file that you want to open.

Figure 3-9
Open File dialog box

▶ Select *demo.mdb* and click Open to continue.

▶ Click OK in the login dialog box.

In the next step, you can specify the tables and variables that you want to import.

Figure 3-10
Select Data step

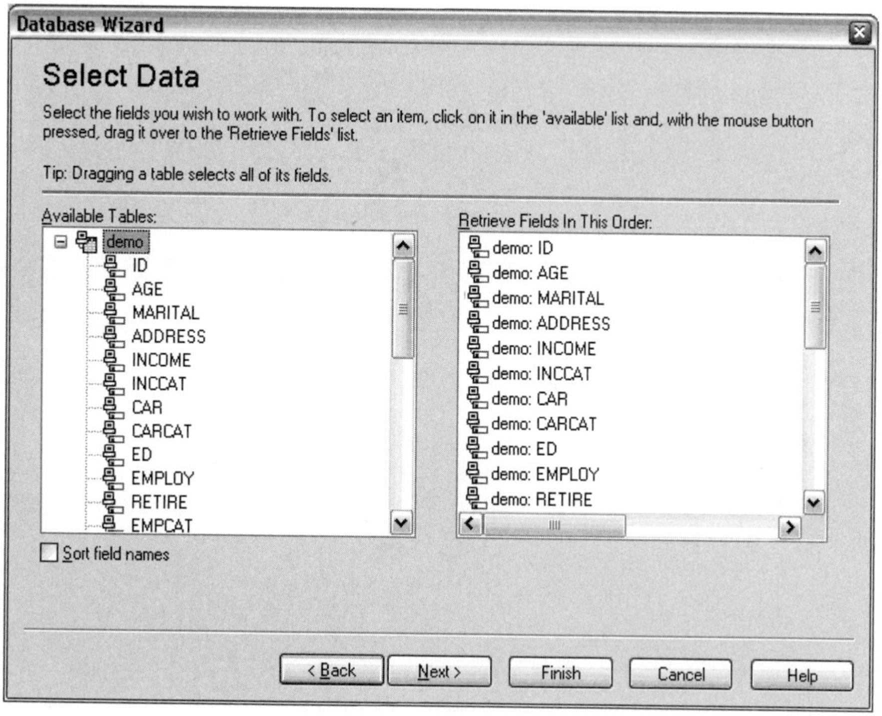

▶ Drag the entire demo table to the Retrieve Fields in This Order list.

▶ Click Next.

In the next step, you select which records (cases) to import.

Figure 3-11
Limit Retrieved Cases step

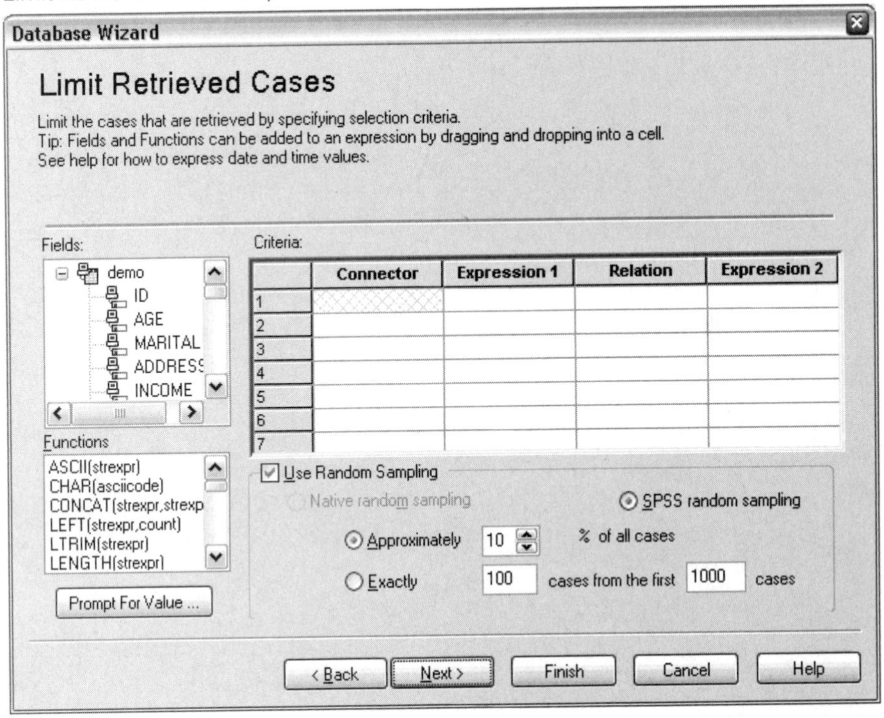

If you do not want to import all cases, you can import a subset of cases (for example, males older than 30), or you can import a random sample of cases from the data source. For large data sources, you may want to limit the number of cases to a small, representative sample to reduce the processing time.

▶ Click Next to continue.

If you are in distributed mode, connected to a remote server (available with SPSS Server), the next step allows you to aggregate the data before reading it into SPSS.

Figure 3-12
Aggregate Data step

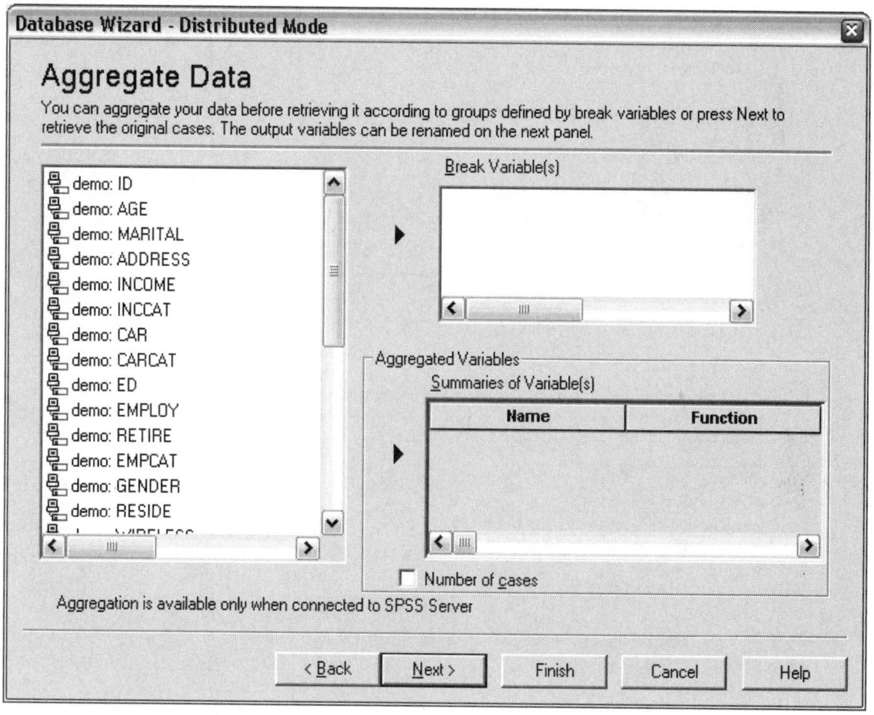

You can also aggregate data after reading it into SPSS, but pre-aggregating may save time for large data sources.

▶ For this example, you don't need to aggregate the data. If you see this step in the Database Wizard, click Next.

Field names are used to create variable names. If necessary, the names are converted to valid variable names. The original field names are preserved as variable labels. You can also change the variable names before importing the database.

Figure 3-13
Define Variables step

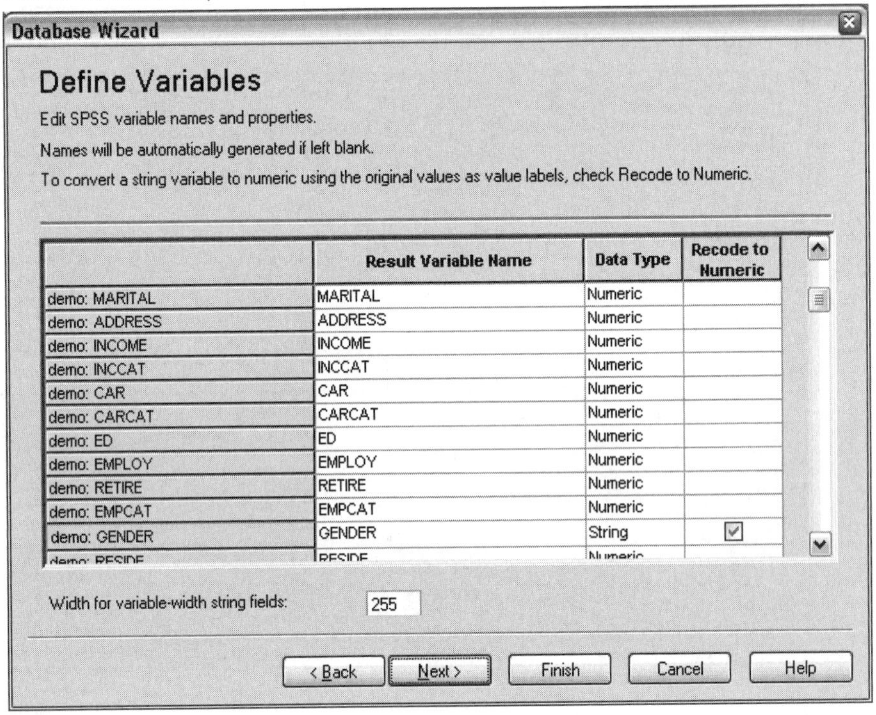

	Result Variable Name	Data Type	Recode to Numeric
demo: MARITAL	MARITAL	Numeric	
demo: ADDRESS	ADDRESS	Numeric	
demo: INCOME	INCOME	Numeric	
demo: INCCAT	INCCAT	Numeric	
demo: CAR	CAR	Numeric	
demo: CARCAT	CARCAT	Numeric	
demo: ED	ED	Numeric	
demo: EMPLOY	EMPLOY	Numeric	
demo: RETIRE	RETIRE	Numeric	
demo: EMPCAT	EMPCAT	Numeric	
demo: GENDER	GENDER	String	☑
demo: RESIDE	RESIDE	Numeric	

Width for variable-width string fields: 255

► Click the Recode to Numeric cell in the Gender field. This option converts string variables to integer variables and retains the original value as the value label for the new variable.

► Click Next to continue.

If you are in distributed mode, connected to a remote server (available with SPSS Server), the next step allows you sort the data before reading it into SPSS.

You can also sort data after reading it into SPSS, but presorting may save time for large data sources.

▶ For this example, you don't need to sort the data. If you see this step in the Database Wizard, click Next.

The SQL statement created from your selections in the Database Wizard appears in the Results step. This statement can be executed now or saved to a file for later use.

Figure 3-14
Results step

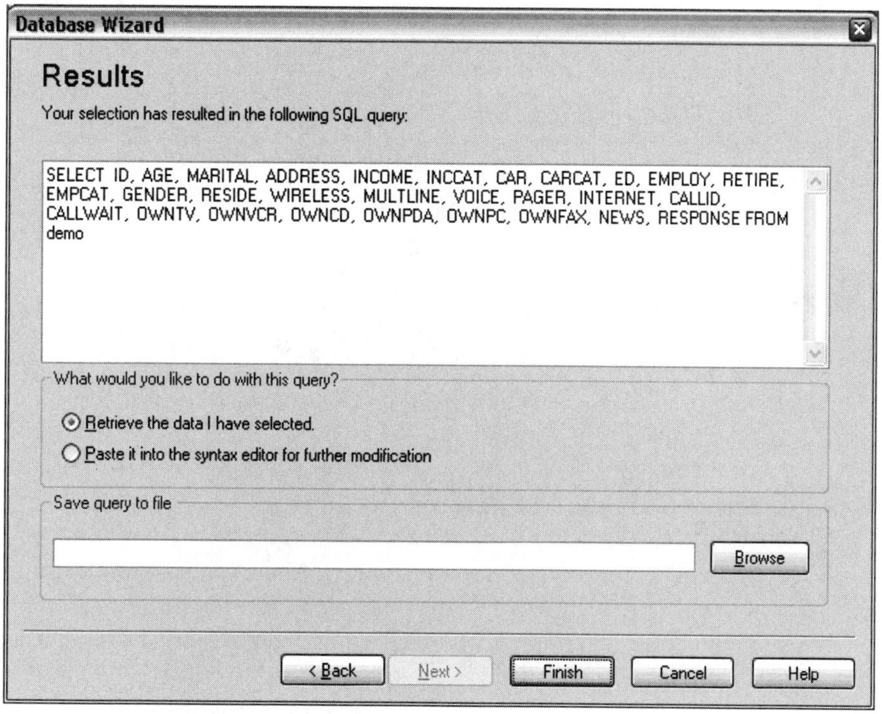

▶ Click Finish to import the data.

All of the data in the Access database that you selected to import are now available in the SPSS Data Editor.

Figure 3-15
Data imported from an Access database

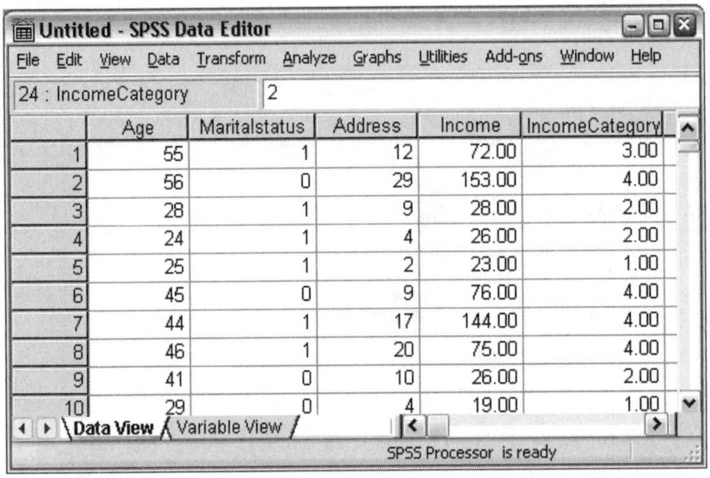

Reading Data from a Text File

Text files are another common source of data. Many spreadsheet programs and databases can save their contents in one of many text file formats. Comma- or tab-delimited files refer to rows of data that use commas or tabs to indicate each variable. In this example, the data are tab delimited.

▶ From the menus choose:
File
   Read Text Data...

▶ Choose Text (*.txt) from the Files of Type list.

Figure 3-16
Open File dialog box

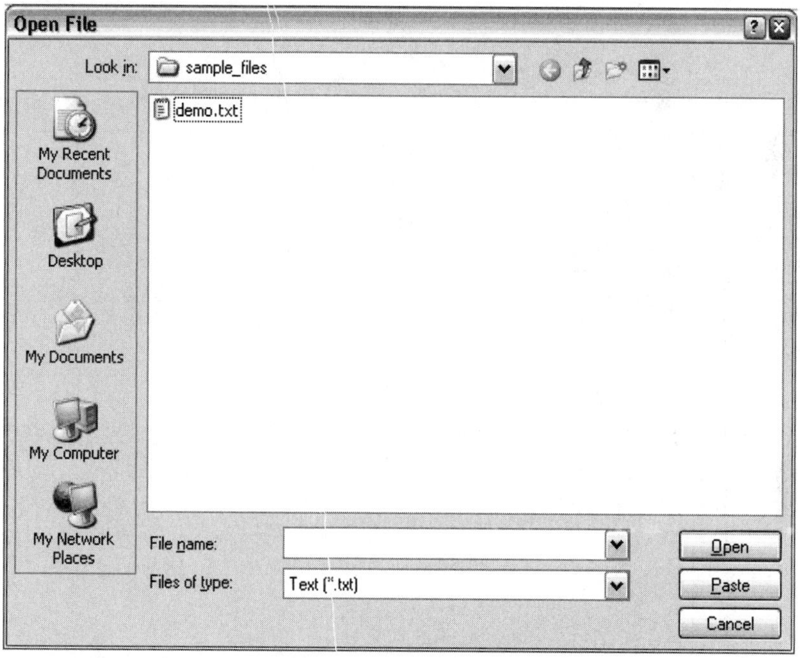

▶ Select *demo.txt* and click Open to read the selected file.

The Text Import Wizard guides you through the process of defining how the specified text file should be interpreted.

Figure 3-17
Text Import Wizard - Step 1 of 6

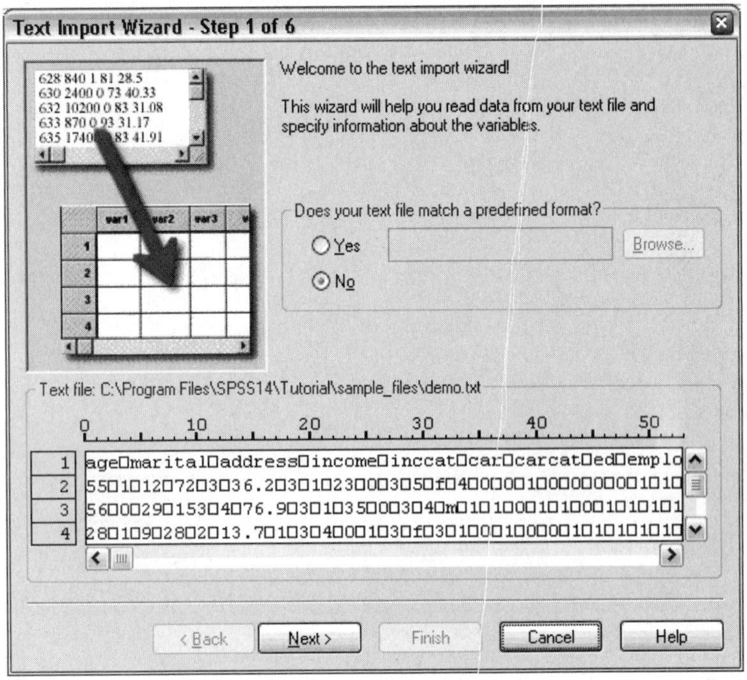

▶ In Step 1, you can choose a predefined format or create a new format in the wizard. Select No to indicate that a new format should be created.

▶ Click Next to continue.

As stated earlier, this file uses tab-delimited formatting. Also, the variable names are defined on the top line of this file.

Figure 3-18
Text Import Wizard - Step 2 of 6

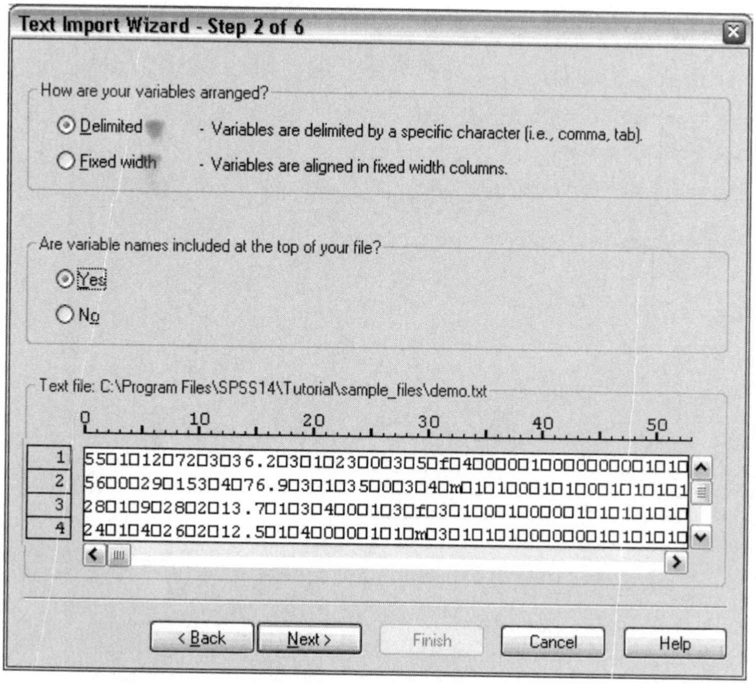

▶ Select Delimited to indicate that the data use a delimited formatting structure.

▶ Select Yes to indicate that variable names should be read from the top of the file.

▶ Click Next to continue.

▶ Type 2 in the top section of next dialog box to indicate that the first row of data starts on the second line of the text file.

Figure 3-19
Text Import Wizard - Step 3 of 6

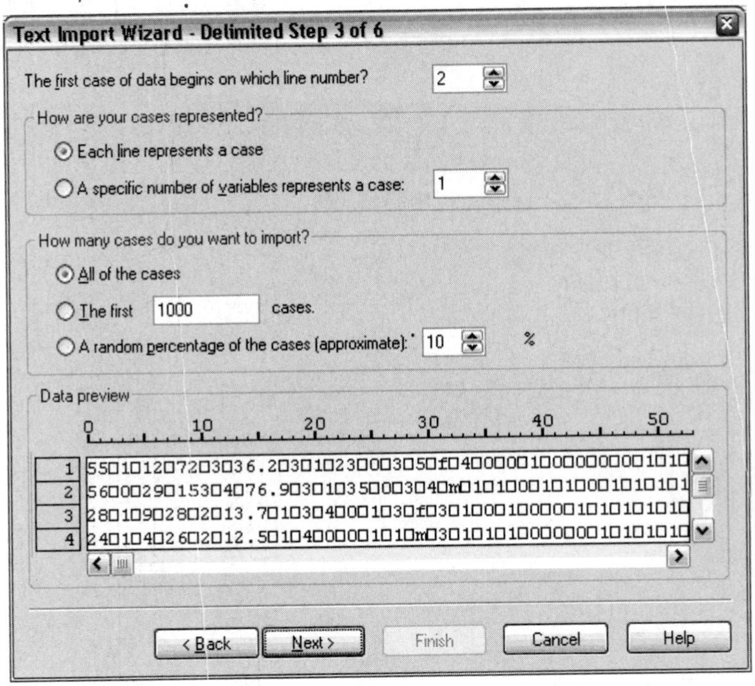

▶ Keep the default values for the remainder of this dialog box, and click Next to continue.

The Data preview in Step 4 provides you with a quick way to ensure that your data are being properly read by SPSS.

Figure 3-20
Text Import Wizard - Step 4 of 6

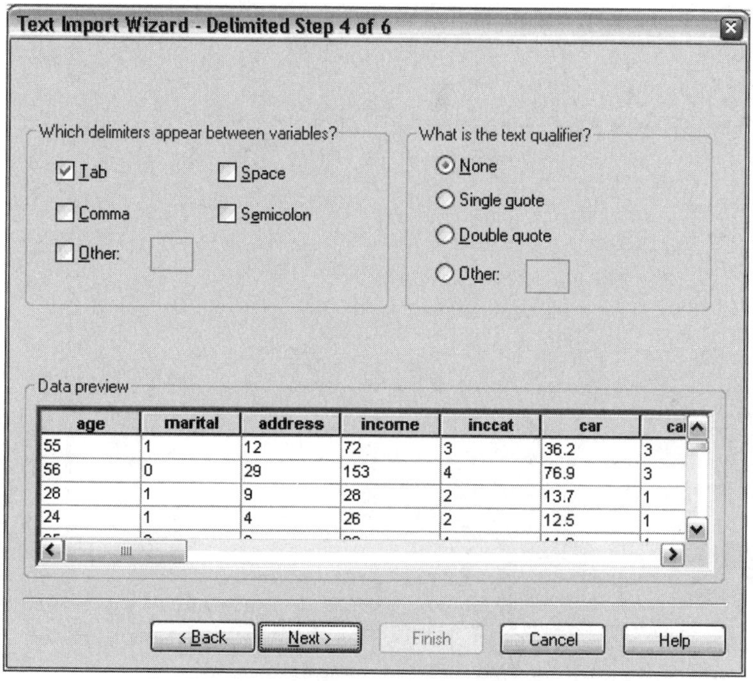

age	marital	address	income	inccat	car	ca
55	1	12	72	3	36.2	3
56	0	29	153	4	76.9	3
28	1	9	28	2	13.7	1
24	1	4	26	2	12.5	1

▶ Select Tab and deselect the other options.

▶ Click Next to continue.

Because the variable names may have been truncated to fit SPSS formatting requirements, this dialog box gives you the opportunity to edit any undesirable names.

Figure 3-21
Text Import Wizard - Step 5 of 6

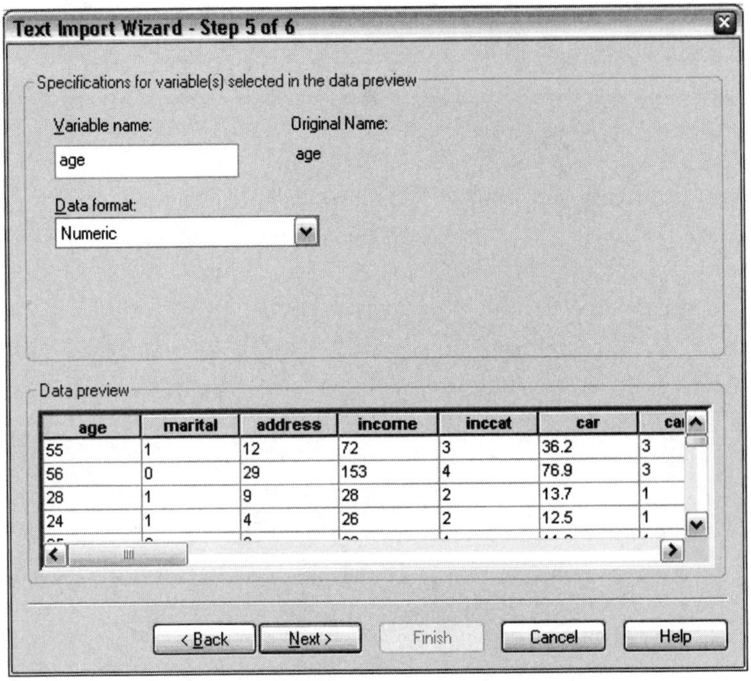

Data types can be defined here as well. For example, it's safe to assume that the income variable is meant to contain a certain dollar amount.

To change a data type:

▶ Under Data preview, select the variable you want to change, which is *Income* in this case.

▶ Select Dollar from the Data format drop-down list.

Figure 3-22
Change the data type

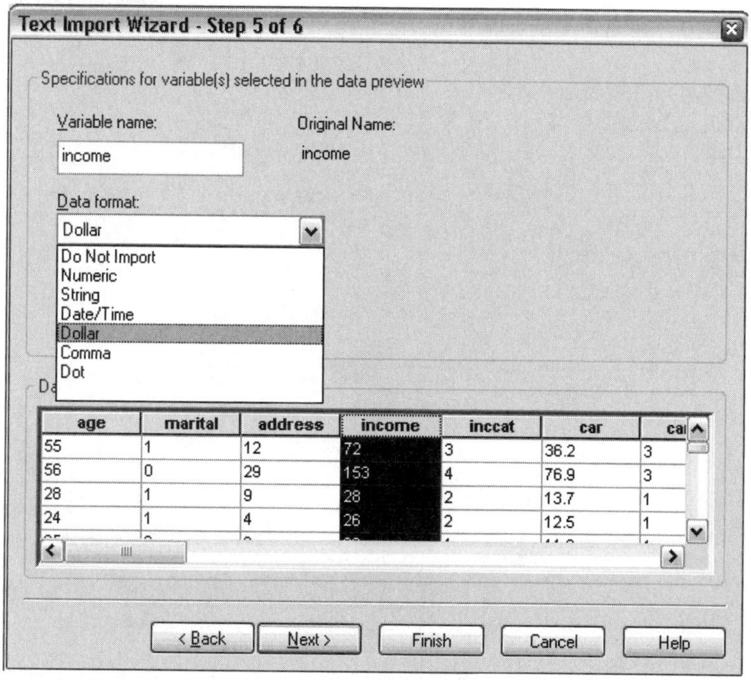

▶ Click Next to continue.

Figure 3-23
Text Import Wizard - Step 6 of 6

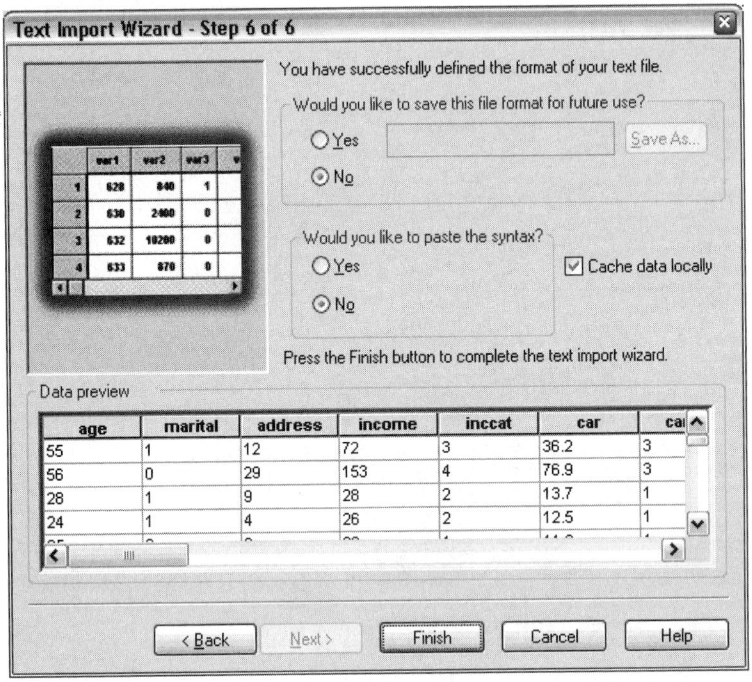

▶ Leave the default selections in this dialog box, and click Finish to import the data.

Saving Data

To save an SPSS data file, the Data Editor window must be the active window.

▶ From the menus choose:
File
  Save

▶ Browse to the desired directory.

▶ Type a name for the file in the File Name text box.

The Variables button can be used to select which variables in the Data Editor are saved to the SPSS data file. By default, all variables in the Data Editor are retained.

▶ Click Save.

The name in the title bar of the Data Editor will change to the filename you specified. This confirms that the file has been successfully saved as an SPSS data file. The file contains both variable information (names, type, and, if provided, labels and missing value codes) and all data values.

Using the Data Editor

The Data Editor displays the contents of the active data file. The information in the Data Editor consists of variables and cases.

- In Data View, columns represent variables, and rows represent cases (observations).

- In Variable View, each row is a variable, and each column is an attribute that is associated with that variable.

Variables are used to represent the different types of data that you have compiled. A common analogy is that of a survey. The response to each question on a survey is equivalent to a variable. Variables come in many different types, including numbers, strings, currency, and dates.

Entering Numeric Data

Data can be entered into the Data Editor, which may be useful for small data files or for making minor edits to larger data files.

▶ Click the Variable View tab at the bottom of the Data Editor window.

You need to define the variables that will be used. In this case, only three variables are needed: *age*, *marital status*, and *income*.

Figure 4-1
Variable names in Variable View

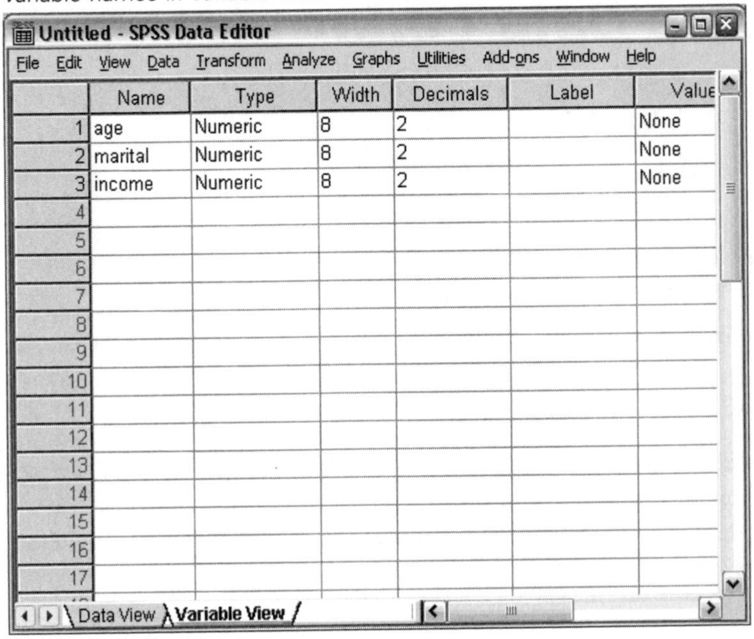

▶ In the first row of the first column, type age.

▶ In the second row, type marital.

▶ In the third row, type income.

New variables are automatically given a Numeric data type.

If you don't enter variable names, unique names are automatically created. However, these names are not descriptive and are not recommended for large data files.

▶ Click the Data View tab to continue entering the data.

The names that you entered in Variable View are now the headings for the first three columns in Data View.

Begin entering data in the first row, starting at the first column.

Figure 4-2
Values entered in Data View

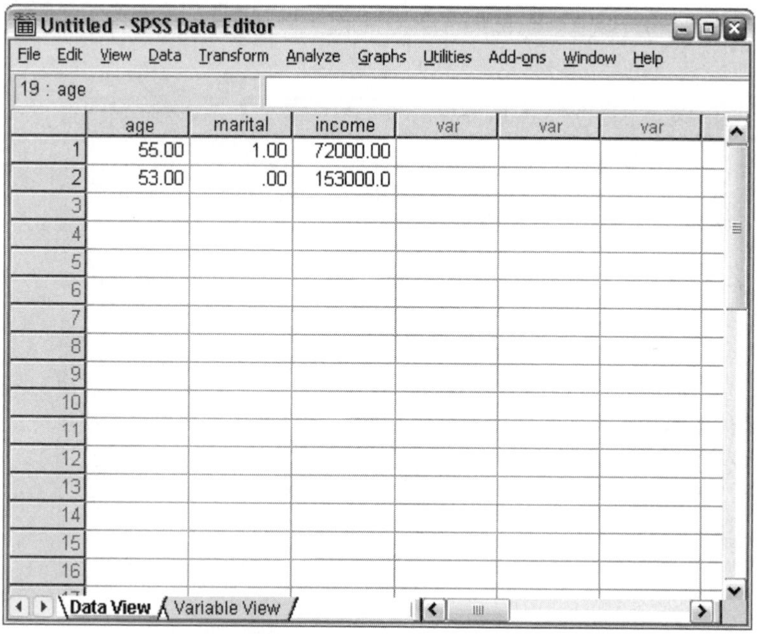

- ▶ In the *age* column, type 55.

- ▶ In the *marital* column, type 1.

- ▶ In the *income* column, type 72000.

- ▶ Move the cursor to the second row of the first column to add the next subject's data.

- ▶ In the *age* column, type 53.

- ▶ In the *marital* column, type 0.

- ▶ In the *income* column, type 153000.

Currently, the *age* and *marital* columns display decimal points, even though their values are intended to be integers. To hide the decimal points in these variables:

- ▶ Click the Variable View tab at the bottom of the Data Editor window.

▶ In the *Decimals* column of the *age* row, type 0 to hide the decimal.

▶ In the *Decimals* column of the *marital* row, type 0 to hide the decimal.

Figure 4-3
Updated decimal property for age and marital

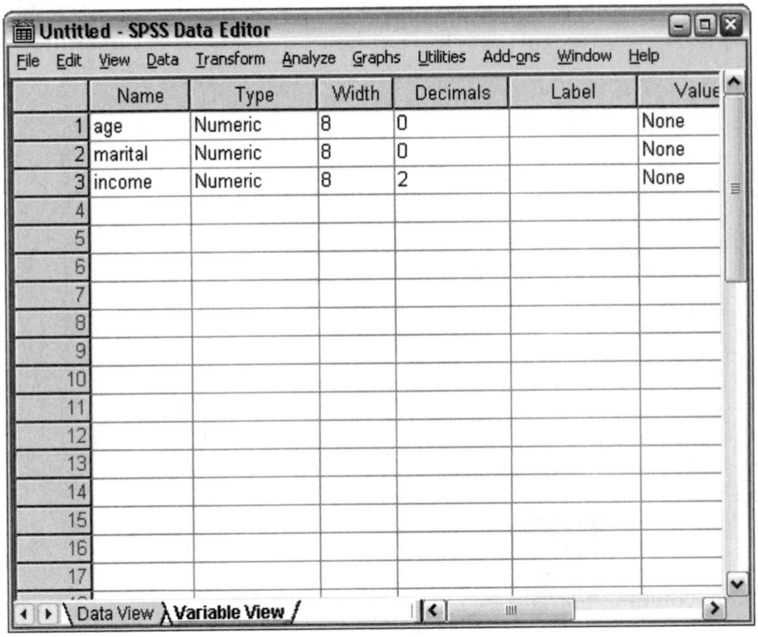

	Name	Type	Width	Decimals	Label	Value
1	age	Numeric	8	0		None
2	marital	Numeric	8	0		None
3	income	Numeric	8	2		None
4						
5						
6						
7						
8						
9						
10						
11						
12						
13						
14						
15						
16						
17						

Data View / Variable View

Entering String Data

Non-numeric data, such as strings of text, can also be entered into the Data Editor.

▶ Click the Variable View tab at the bottom of the Data Editor window.

▶ In the first cell of the first empty row, type **sex** for the variable name.

▶ Click the *Type* cell next to your entry.

▶ Click the button on the right side of the *Type* cell to open the Variable Type dialog box.

Figure 4-4
Button shown in Type cell for sex

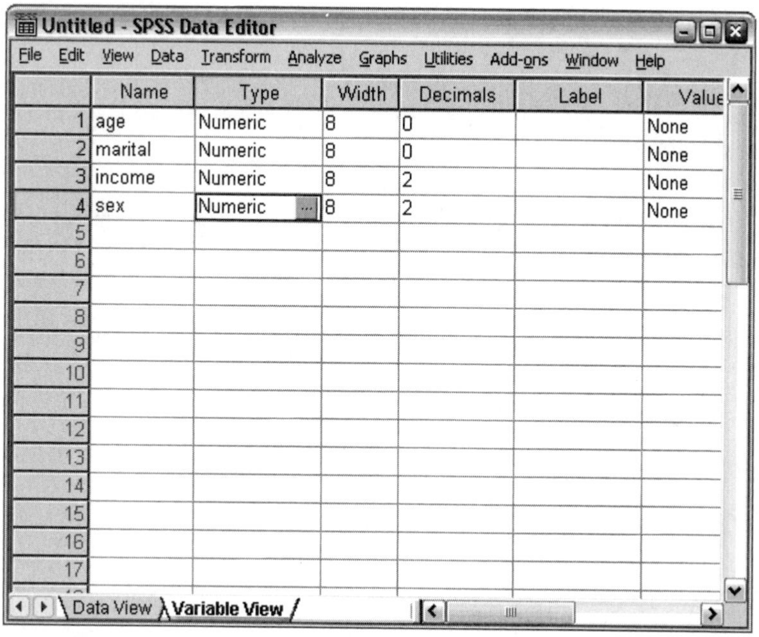

		Name	Type	Width	Decimals	Label	Value
1	age	Numeric	8	0		None
2	marital	Numeric	8	0		None
3	income	Numeric	8	2		None
4	sex	Numeric ...	8	2		None

Untitled - SPSS Data Editor
File Edit View Data Transform Analyze Graphs Utilities Add-ons Window Help

Data View \ Variable View /

▶ Select String to specify the variable type.

▶ Click OK to save your selection and return to the Data Editor.

Figure 4-5
Variable Type dialog box

Variable Type

○ Numeric
○ Comma
○ Dot
○ Scientific notation   Characters: 8
○ Date
○ Dollar
○ Custom currency
⊙ String

OK
Cancel
Help

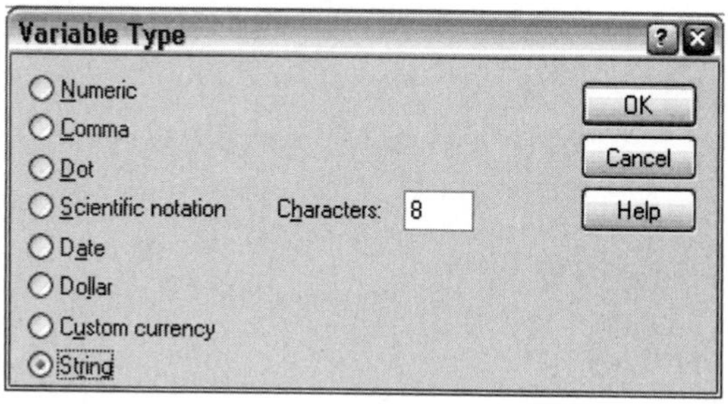

Defining Data

In addition to defining data types, you can also define descriptive variable labels and value labels for variable names and data values. These descriptive labels are used in statistical reports and charts.

Adding Variable Labels

Labels are meant to provide descriptions of variables. These descriptions are often longer versions of variable names. Labels can be up to 255 bytes. These labels are used in your output to identify the different variables.

▶ Click the Variable View tab at the bottom of the Data Editor window.

▶ In the *Label* column of the *age* row, type Respondent's Age.

▶ In the *Label* column of the *marital* row, type Marital Status.

▶ In the *Label* column of the *income* row, type Household Income.

▶ In the *Label* column of the *sex* row, type Gender.

Figure 4-6
Variable labels entered in Variable View

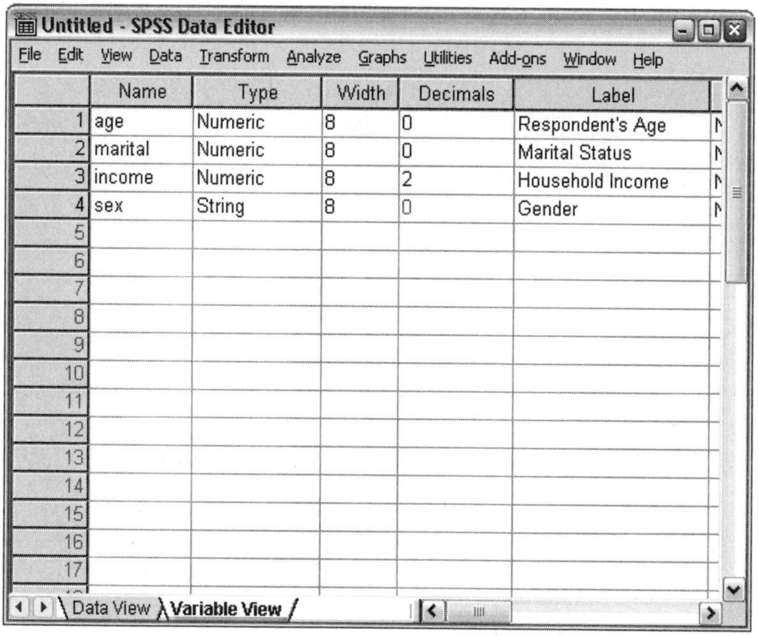

	Name	Type	Width	Decimals	Label	
1	age	Numeric	8	0	Respondent's Age	
2	marital	Numeric	8	0	Marital Status	
3	income	Numeric	8	2	Household Income	
4	sex	String	8	0	Gender	
5						
6						
7						
8						
9						
10						
11						
12						
13						
14						
15						
16						
17						

Changing Variable Type and Format

The *Type* column displays the current data type for each variable. The most common data types are numeric and string, but many other formats are supported. In the current data file, the *income* variable is defined as a numeric type.

▶ Click the *Type* cell for the *income* row, and then click the button on the right side of the cell to open the Variable Type dialog box.

▶ Select Dollar.

Figure 4-7
Variable Type dialog box

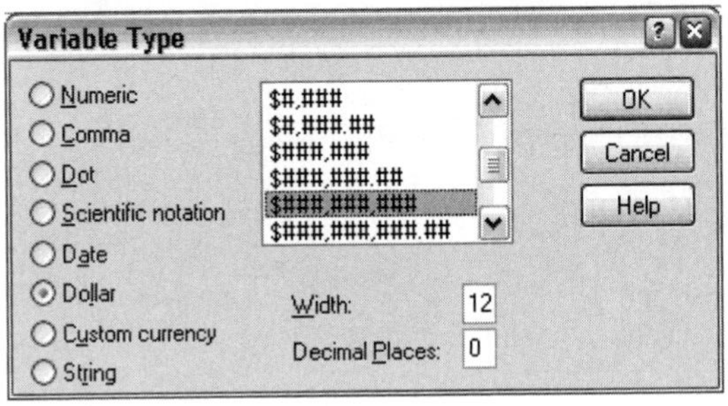

The formatting options for the currently selected data type are displayed.

▶ For the format of the currency in this example, select $###,###,###.

▶ Click OK to save your changes.

Adding Value Labels for Numeric Variables

Value labels provide a method for mapping your variable values to a string label. In this example, there are two acceptable values for the *marital* variable. A value of 0 means that the subject is single, and a value of 1 means that he or she is married.

▶ Click the *Values* cell for the *marital* row, and then click the button on the right side of the cell to open the Value Labels dialog box.

The **value** is the actual numeric value.

The **value label** is the string label that is applied to the specified numeric value.

▶ Type 0 in the Value field.

▶ Type Single in the Label field.

▶ Click Add to add this label to the list.

Figure 4-8
Value Labels dialog box

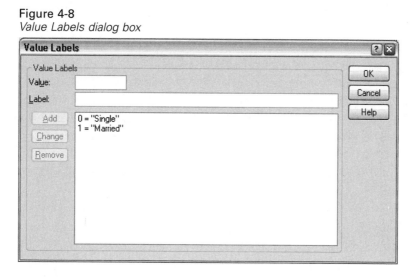

▶ Type 1 in the Value field, and type Married in the Label field.

▶ Click Add, and then click OK to save your changes and return to the Data Editor.

These labels can also be displayed in Data View, which can make your data more readable.

▶ Click the Data View tab at the bottom of the Data Editor window.

▶ From the menus choose:
View
  Value Labels

The labels are now displayed in a list when you enter values in the Data Editor. This setup has the benefit of suggesting a valid response and providing a more descriptive answer.

If the Value Labels menu item is already active (with a check mark next to it), choosing Value Labels again will turn *off* the display of value labels.

Figure 4-9
Value labels displayed in Data View

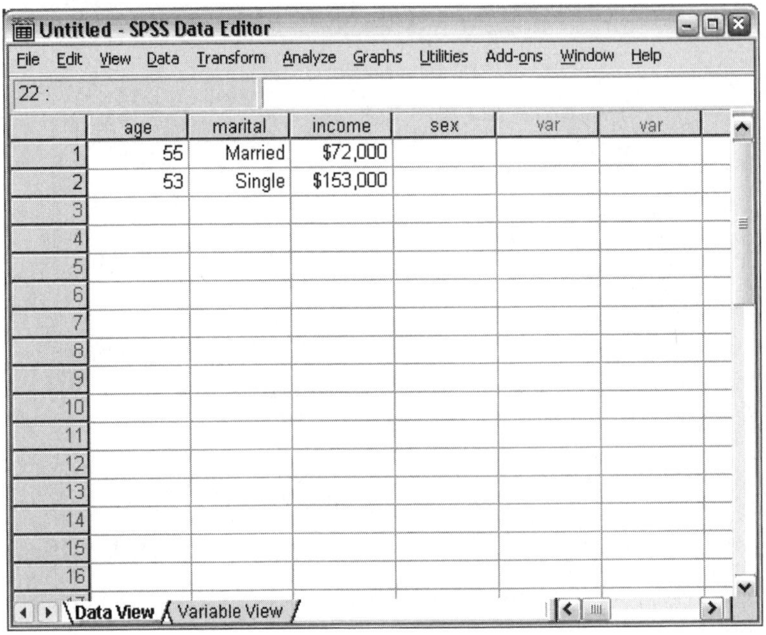

Adding Value Labels for String Variables

String variables may require value labels as well. For example, your data may use single letters, *M* or *F*, to identify the sex of the subject. Value labels can be used to specify that *M* stands for *Male* and *F* stands for *Female*.

▶ Click the Variable View tab at the bottom of the Data Editor window.

▶ Click the *Values* cell in the *sex* row, and then click the button on the right side of the cell to open the Value Labels dialog box.

▶ Type F in the Value field, and then type Female in the Label field.

▶ Click Add to add this label to your data file.

Figure 4-10
Value Labels dialog box

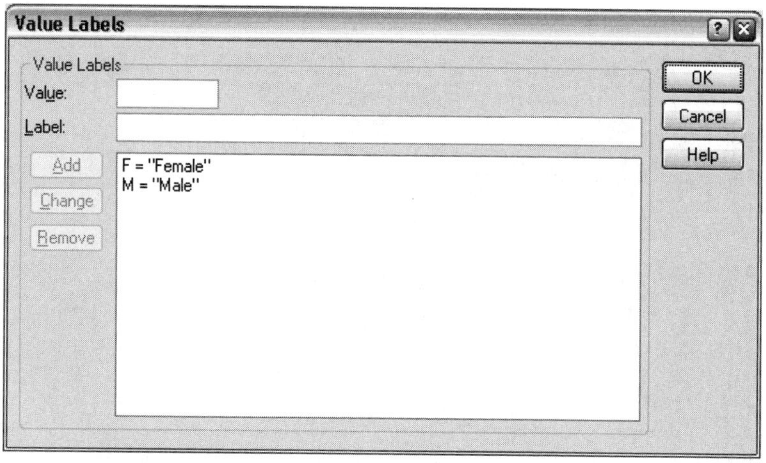

▶ Type M in the Value field, and type Male in the Label field.

▶ Click Add, and then click OK to save your changes and return to the Data Editor.

Because string values are case-sensitive, you should be consistent. A lowercase *m* is not the same as an uppercase *M*.

Using Value Labels for Data Entry

You can use value labels for data entry.

▶ Click the Data View tab at the bottom of the Data Editor window.

▶ In the first row, select the cell for *sex*.

▶ Click the button on the right side of the cell, and then choose Male from the drop-down list.

▶ In the second row, select the cell for *sex*.

▶ Click the button on the right side of the cell, and then choose Female from the drop-down list.

Figure 4-11
Using variable labels to select values

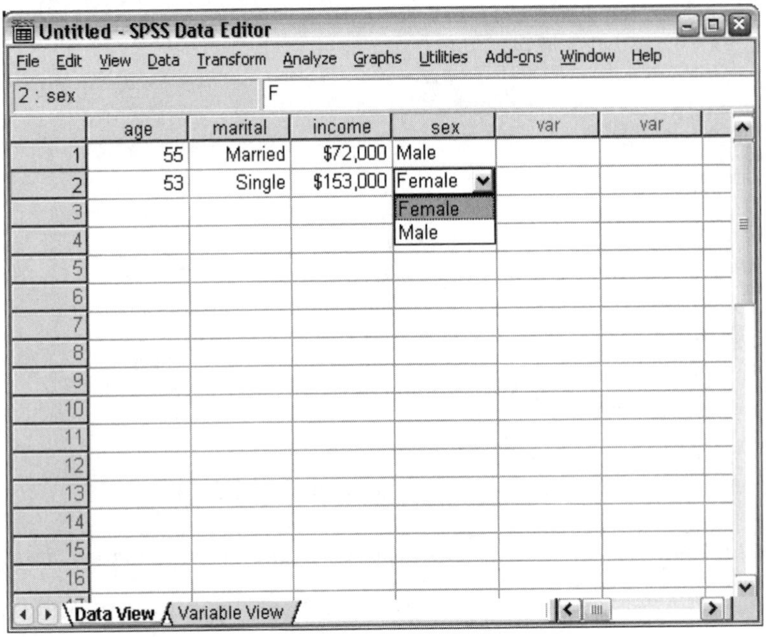

Only defined values are listed, which ensures that the entered data are in a format that you expect.

Handling Missing Data

Missing or invalid data are generally too common to ignore. Survey respondents may refuse to answer certain questions, may not know the answer, or may answer in an unexpected format. If you don't filter or identify these data, your analysis may not provide accurate results.

For numeric data, empty data fields or fields containing invalid entries are converted to system-missing, which is identifiable by a single period.

Figure 4-12
Missing values displayed as periods

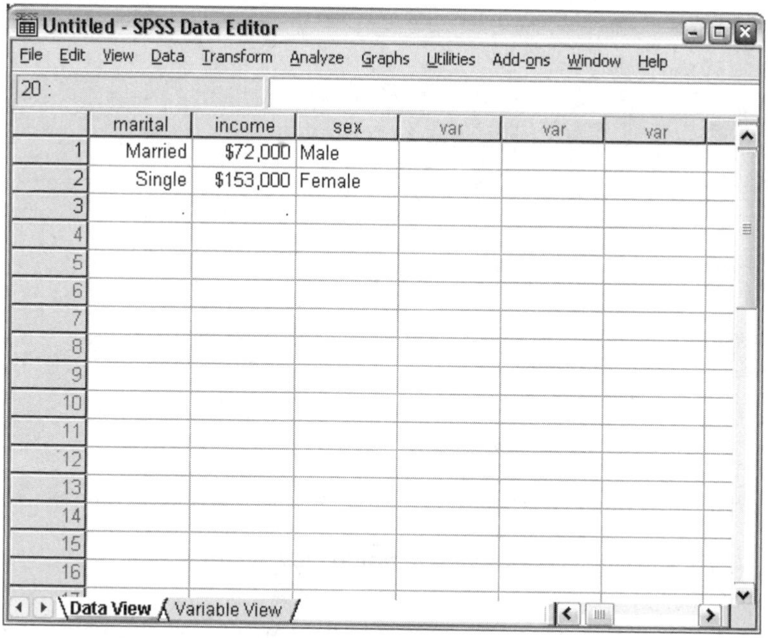

The reason a value is missing may be important to your analysis. For example, you may find it useful to distinguish between those respondents who refused to answer a question and those respondents who didn't answer a question because it was not applicable.

Missing Values for a Numeric Variable

▶ Click the Variable View tab at the bottom of the Data Editor window.

▶ Click the *Missing* cell in the *age* row, and then click the button on the right side of the cell to open the Missing Values dialog box.

In this dialog box, you can specify up to three distinct missing values, or you can specify a range of values plus one additional discrete value.

Figure 4-13
Missing Values dialog box

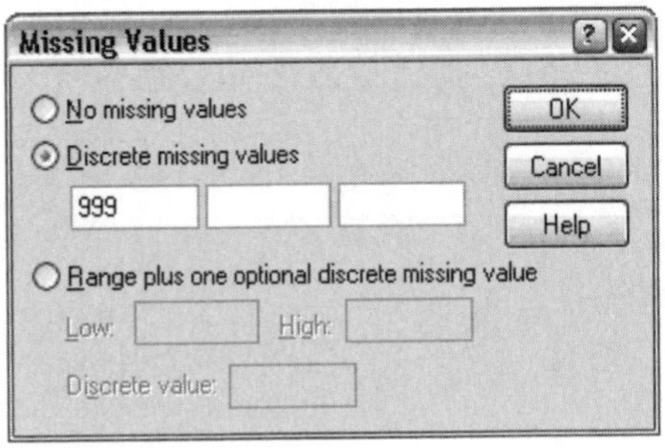

▶ Select Discrete missing values.

▶ Type **999** in the first text box and leave the other two text boxes empty.

▶ Click **OK** to save your changes and return to the Data Editor.

Now that the missing data value has been added, a label can be applied to that value.

▶ Click the *Values* cell in the *age* row, and then click the button on the right side of the cell to open the Value Labels dialog box.

▶ Type **999** in the Value field.

▶ Type No Response in the Label field.

Figure 4-14
Value Labels dialog box

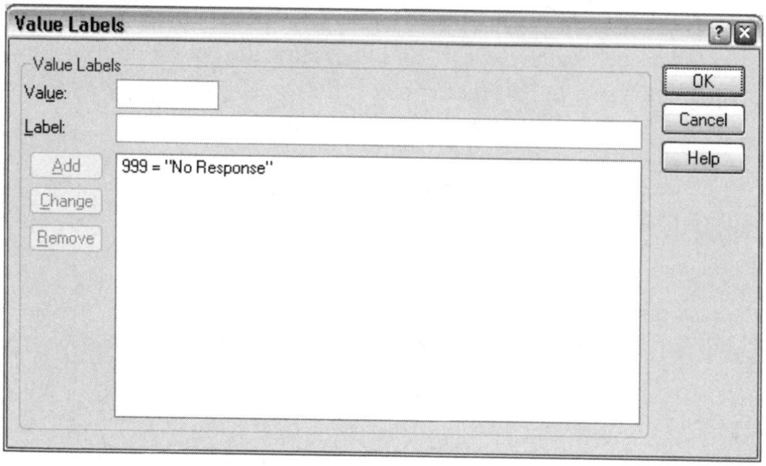

▶ Click Add to add this label to your data file.

▶ Click OK to save your changes and return to the Data Editor.

Missing Values for a String Variable

Missing values for string variables are handled similarly to the missing values for numeric variables. However, unlike numeric variables, empty fields in string variables are not designated as system-missing. Rather, they are interpreted as an empty string.

▶ Click the Variable View tab at the bottom of the Data Editor window.

▶ Click the *Missing* cell in the *sex* row, and then click the button on the right side of the cell to open the Missing Values dialog box.

▶ Select Discrete missing values.

▶ Type NR in the first text box.

Missing values for string variables are case-sensitive. So, a value of *nr* is not treated as a missing value.

▶ Click OK to save your changes and return to the Data Editor.

Now you can add a label for the missing value.

▶ Click the *Values* cell in the *sex* row, and then click the button on the right side of the cell to open the Value Labels dialog box.

▶ Type NR in the Value field.

▶ Type No Response in the Label field.

Figure 4-15
Value Labels dialog box

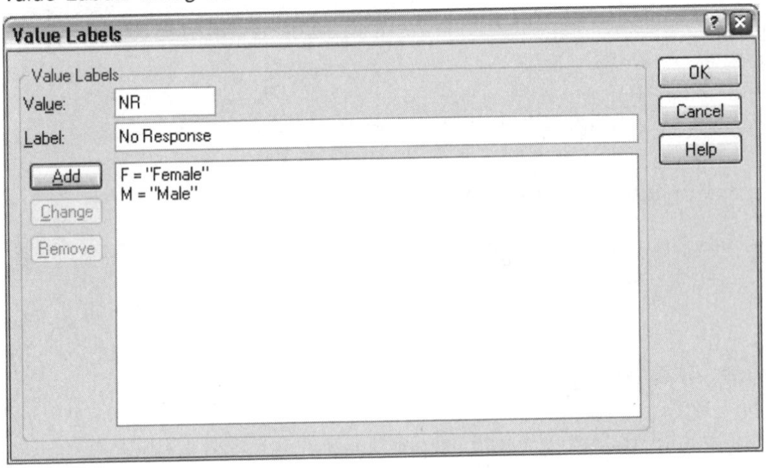

▶ Click Add to add this label to your project.

▶ Click OK to save your changes and return to the Data Editor.

Copying and Pasting Variable Attributes

After you've defined variable attributes for a variable, you can copy these attributes and apply them to other variables.

▶ In Variable View, type **agewed** in the first cell of the first empty row.

Figure 4-16
agewed variable in Variable View

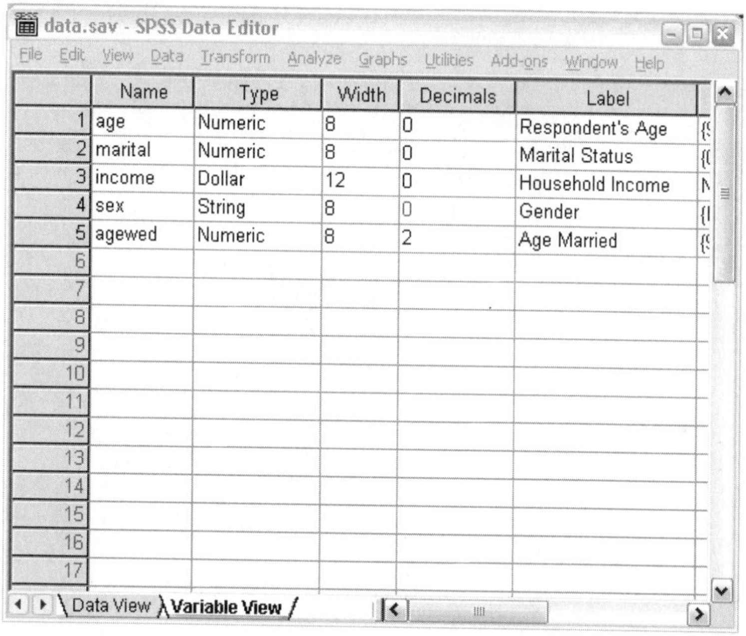

	Name	Type	Width	Decimals	Label	
1	age	Numeric	8	0	Respondent's Age	{}
2	marital	Numeric	8	0	Marital Status	{(
3	income	Dollar	12	0	Household Income	N
4	sex	String	8	0	Gender	{(
5	agewed	Numeric	8	2	Age Married	{}
6						
7						
8						
9						
10						
11						
12						
13						
14						
15						
16						
17						

data.sav - SPSS Data Editor

File Edit View Data Transform Analyze Graphs Utilities Add-ons Window Help

\ Data View \Variable View /

▶ In the *Label* column, type **Age Married**.

▶ Click the *Values* cell in the *age* row.

▶ From the menus choose:
Edit
   Copy

▶ Click the *Values* cell in the *agewed* row.

▶ From the menus choose:
Edit
   Paste

The defined values from the *age* variable are now applied to the *agewed* variable.

To apply the attribute to multiple variables, simply select multiple target cells (click and drag down the column).

Figure 4-17
Multiple cells selected

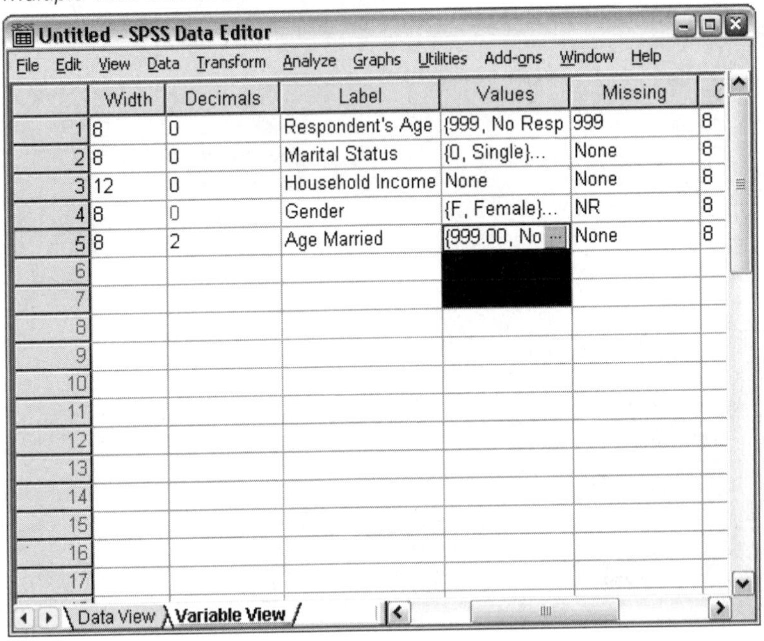

		Width	Decimals	Label	Values	Missing	C
	1	8	0	Respondent's Age	{999, No Resp	999	8
	2	8	0	Marital Status	{0, Single}...	None	8
	3	12	0	Household Income	None	None	8
	4	8	0	Gender	{F, Female}...	NR	8
	5	8	2	Age Married	{999.00, No ...	None	8
	6						
	7						
	8						
	9						
	10						
	11						
	12						
	13						
	14						
	15						
	16						
	17						

Untitled - SPSS Data Editor

File Edit View Data Transform Analyze Graphs Utilities Add-ons Window Help

Data View \ Variable View /

When you paste the attribute, it is applied to all of the selected cells.

New variables are automatically created if you paste the values into empty rows.

To copy all attributes from one variable to another variable:

▶ Click the row number in the *marital* row.

Figure 4-18
Selected row

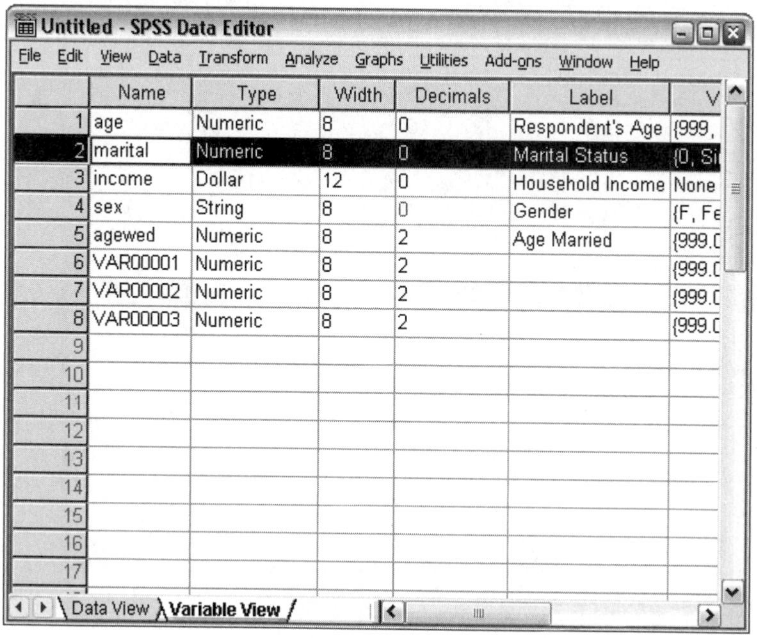

▶ From the menus choose:
Edit
  Copy

▶ Click the row number of the first empty row.

▶ From the menus choose:
Edit
  Paste

All attributes of the *marital* variable are applied to the new variable.

Figure 4-19
All values pasted into a row

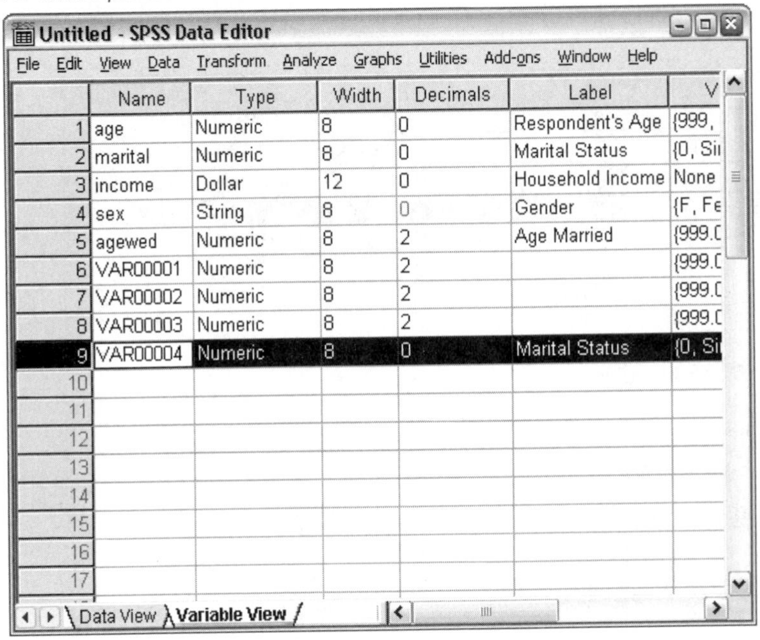

Defining Variable Properties for Categorical Variables

For categorical (nominal, ordinal) data, you can define value labels and other variable properties. The Define Variable Properties process:

■ Scans the actual data values and lists all unique data values for each selected variable.

■ Identifies unlabeled values and provides an "auto-label" feature.

■ Provides the ability to copy defined value labels from another variable to the selected variable or from the selected variable to additional variables.

This example uses the data file *demo.sav*. This data file already has defined value labels, so we will enter a value for which there is no defined value label.

▶ In Data View of the Data Editor, click the first data cell for the variable *ownpc* (you may have to scroll to the right), and then enter 99.

▶ From the menus choose:

Data
  Define Variable Properties...

Figure 4-20
Initial Define Variable Properties dialog box

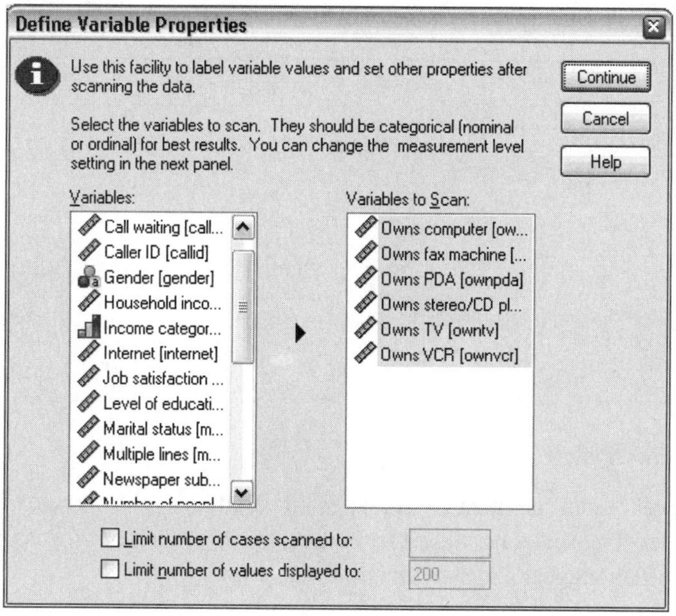

In the initial Define Variable Properties dialog box, you select the nominal or ordinal variables for which you want to define value labels and/or other properties.

▶ Drag and drop *Owns computer [ownpc]* through *Owns VCR [ownvcr]* into the Variables to Scan list.

You might notice that the measurement level icons for all of the selected variables indicate that they are scale variables, not categorical variables. All of the selected variables in this example are really categorical variables that use the numeric values 0 and 1 to stand for *No* and *Yes*, respectively—and one of the variable properties that we'll change with Define Variable Properties is the measurement level.

▶ Click Continue.

Figure 4-21
Define Variable Properties main dialog box

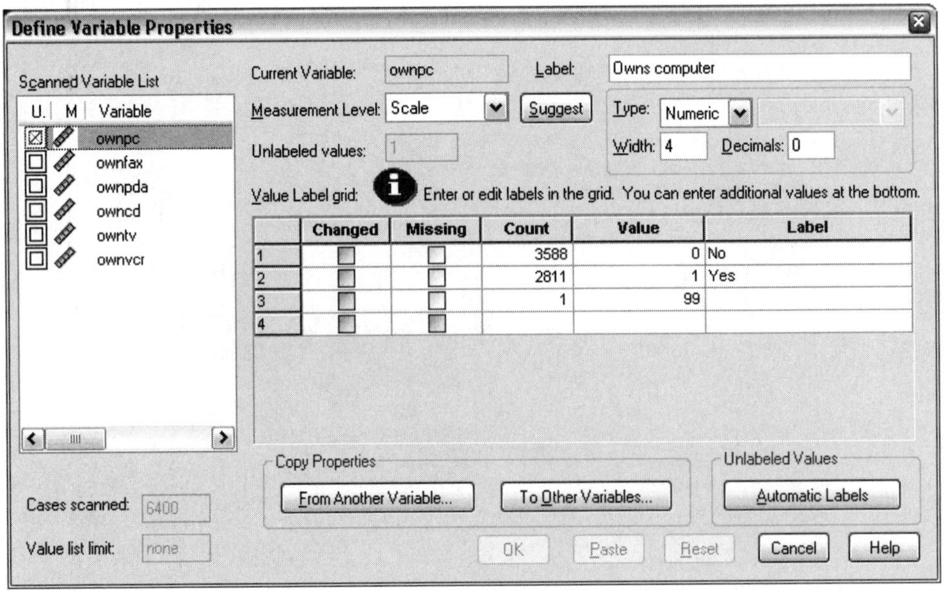

▶ In the Scanned Variable List, select *ownpc*.

The current level of measurement for the selected variable is scale. You can change the measurement level by selecting a level from the drop-down list, or you can let Define Variable Properties suggest a measurement level.

▶ Click Suggest.

The Suggest Measurement Level dialog box is displayed.

Figure 4-22
Suggest Measurement Level dialog box

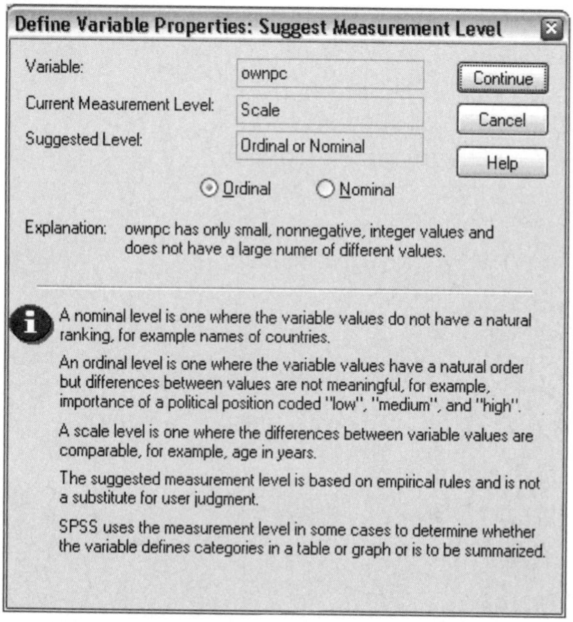

Because the variable doesn't have very many different values and all of the scanned cases contain integer values, the proper measurement level is probably ordinal or nominal.

▶ Select Ordinal, and then click Continue.

The measurement level for the selected variable is now ordinal.

The Value Label grid displays all of the unique data values for the selected variable, any defined value labels for these values, and the number of times (count) that each value occurs in the scanned cases.

The value that we entered in Data View, 99, is displayed in the grid. The count is only 1 because we changed the value for only one case, and the *Label* column is empty because we haven't defined a value label for 99 yet. An X in the first column of the Scanned Variable List also indicates that the selected variable has at least one observed value without a defined value label.

▶ In the *Label* column for the value of 99, enter No answer.

▶ Check the box in the *Missing* column for the value 99 to identify the value 99 as **user-missing**.

Data values that are specified as user-missing are flagged for special treatment and are excluded from most calculations.

Figure 4-23
New variable properties defined for ownpc

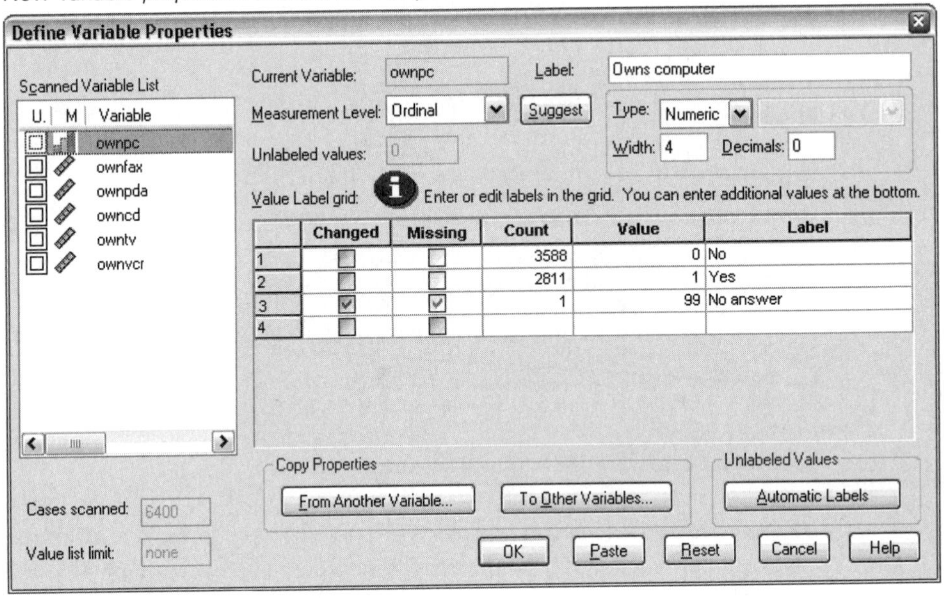

Before we complete the job of modifying the variable properties for *ownpc*, let's apply the same measurement level, value labels, and missing values definitions to the other variables in the list.

▶ In the Copy Properties area, click To Other Variables.

Figure 4-24
Apply Labels and Level dialog box

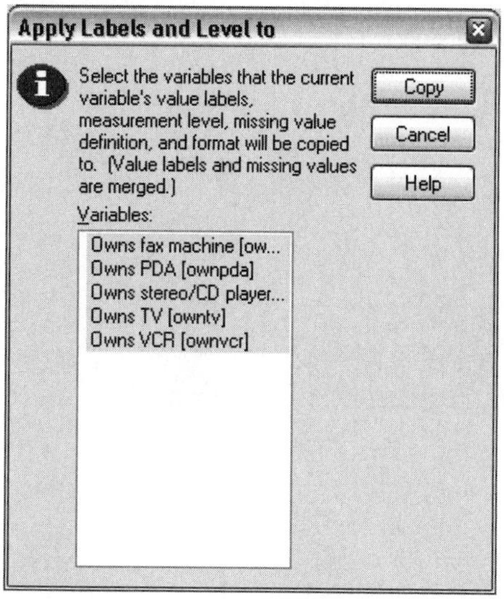

In the Apply Labels and Level dialog box, select all of the variables in the list, and then click Copy.

If you select any other variable in the Scanned Variable List of the Define Variable Properties main dialog box now, you'll see that they are all ordinal variables, with a value of 99 defined as user-missing and a value label of *No answer*.

Figure 4-25
New variable properties defined for ownfax

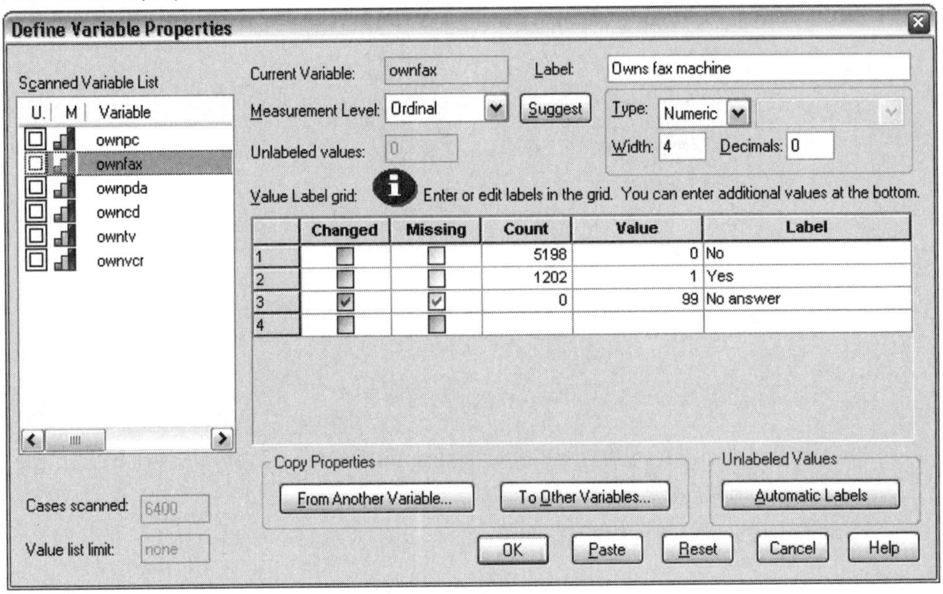

► Click OK to save all of the variable properties that you have defined.

Working with Multiple Data Sources

Starting with SPSS 14.0, SPSS can have multiple data sources open at the same time, making it easier to:

- Switch back and forth between data sources.

- Compare the contents of different data sources.

- Copy and paste data between data sources.

- Create multiple subsets of cases and/or variables for analysis.

- Merge multiple data sources from various data formats (for example, spreadsheet, database, text data) without saving each data source in SPSS format first.

Basic Handling of Multiple Data Sources

Figure 5-1
Two data sources open at same time

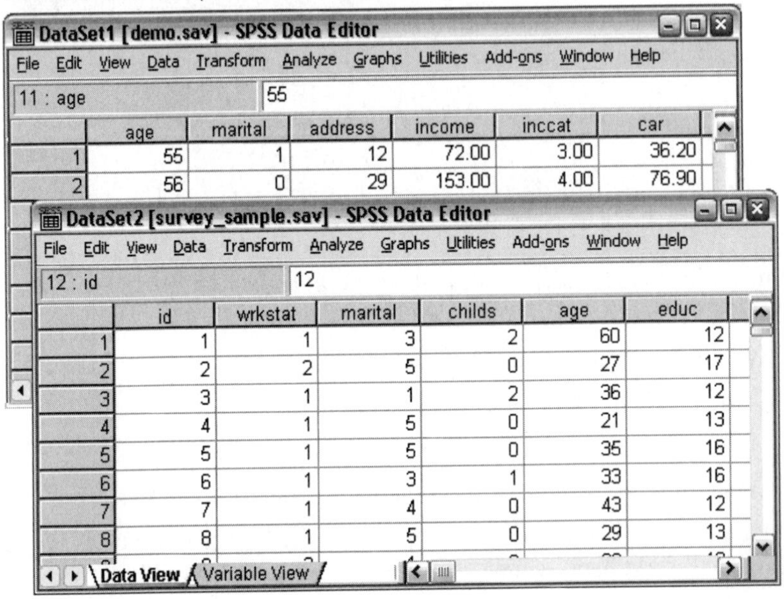

Each data source that you open is displayed in a new Data Editor window.

- Any previously open data sources remain open and available for further use.

- When you first open a data source, it automatically becomes the **active dataset**.

- You can change the active dataset simply by clicking anywhere in the Data Editor window of the data source that you want to use or by selecting the Data Editor window for that data source from the Window menu.

■ Only the variables in the active dataset are available for analysis.

Figure 5-2
Variable list containing variables in the active dataset

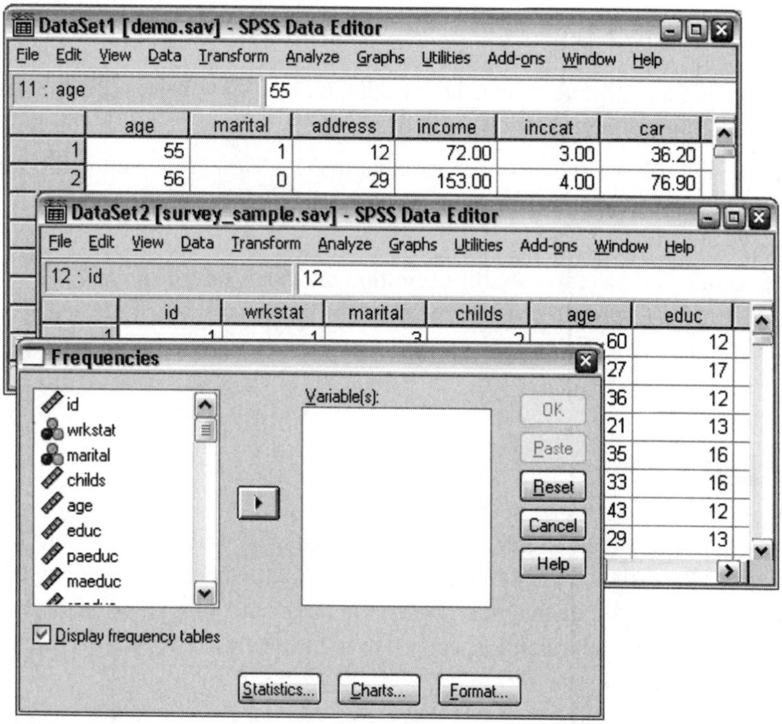

■ You cannot change the active dataset when any dialog box that accesses the data is open (including all dialog boxes that display variable lists).

■ At least one Data Editor window must be open during a session. When you close the last open Data Editor window, SPSS automatically shuts down, prompting you to save changes first.

Note: If you use command syntax to open data sources (for example, GET FILE, GET DATA), you need to explicitly name each dataset in order to have more than one data source open at the same time.

Copying and Pasting Information between Datasets

You can copy both data and variable definition attributes from one dataset to another dataset in basically the same way that you copy and paste information within a single data file.

- Copying and pasting selected data cells in Data View pastes just the data values, with no variable definition attributes.

- Copying and pasting an entire variable in Data View by selecting the variable name at the top of the column pastes all of the data and all of the variable definition attributes for that variable.

- Copying and pasting variable definition attributes or entire variables in Variable View pastes the selected attributes (or the entire variable definition) but does not paste any data values.

Renaming Datasets

When you open a data source through the menus and dialogs, each data source is automatically assigned a dataset name of *DataSetn*, where *n* is a sequential integer value, and when you open a data source via command syntax, no dataset name is assigned unless you explicitly specify one with DATASET NAME. To provide more descriptive dataset names:

▶ From the menus in the Data Editor window for the dataset for which you want to change the dataset name, choose:

File
  Rename Dataset

▶ Enter a new dataset name that conforms to SPSS variable naming rules.

Examining Summary Statistics for Individual Variables

This chapter discusses simple summary measures and how the level of measurement of a variable influences the types of statistics that should be used. We will use the data file *demo.sav*.

Level of Measurement

Different summary measures are appropriate for different types of data, depending on the level of measurement:

Categorical. Data with a limited number of distinct values or categories (for example, gender or marital status). Also referred to as **qualitative data**. Categorical variables can be string (alphanumeric) data or numeric variables that use numeric codes to represent categories (for example, 0 = *Unmarried* and 1 = *Married*). There are two basic types of categorical data:

■ **Nominal.** Categorical data where there is no inherent order to the categories. For example, a job category of *sales* is not higher or lower than a job category of *marketing* or *research*.

■ **Ordinal.** Categorical data where there is a meaningful order of categories, but there is not a measurable distance between categories. For example, there is an order to the values *high*, *medium*, and *low*, but the "distance" between the values cannot be calculated.

Scale. Data measured on an interval or ratio scale, where the data values indicate both the order of values and the distance between values. For example, a salary of $72,195 is higher than a salary of $52,398, and the distance between the two values is $19,797. Also referred to as **quantitative** or **continuous data**.

Summary Measures for Categorical Data

For categorical data, the most typical summary measure is the number or percentage of cases in each category. The **mode** is the category with the greatest number of cases. For ordinal data, the **median** (the value at which half of the cases fall above and below) may also be a useful summary measure if there is a large number of categories.

The Frequencies procedure produces frequency tables that display both the number and percentage of cases for each observed value of a variable.

▶ From the menus choose:

Analyze
  Descriptive Statistics
    Frequencies...

▶ Select *Owns PDA [ownpda]* and *Owns TV [owntv]* and move them into the Variable(s) list.

Figure 6-1
Categorical variables selected for analysis

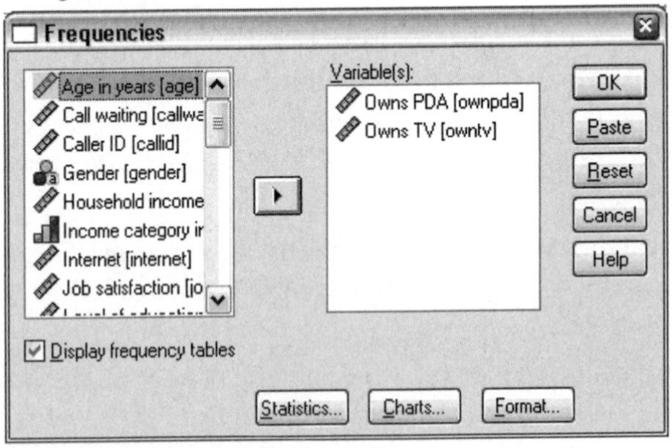

▶ Click OK to run the procedure.

Figure 6-2
Frequency tables

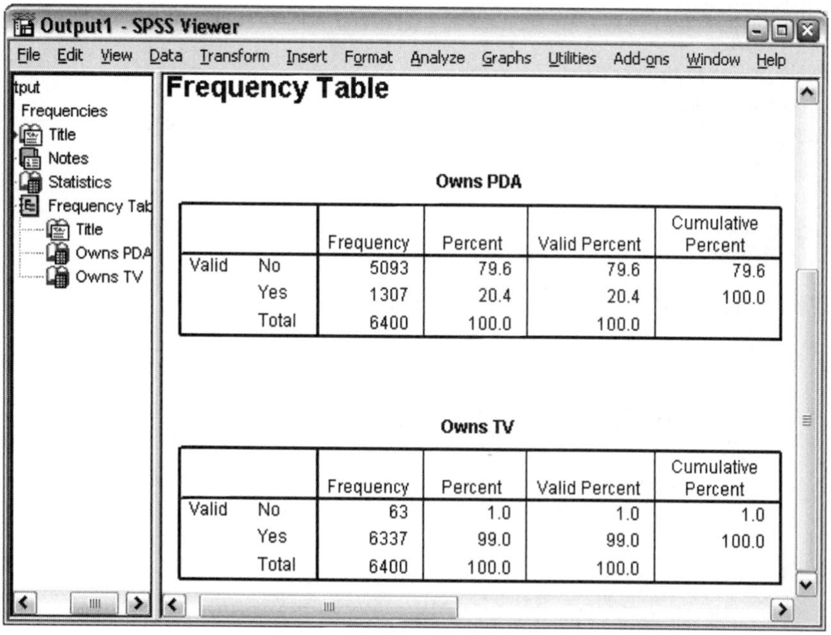

Owns PDA

		Frequency	Percent	Valid Percent	Cumulative Percent
Valid	No	5093	79.6	79.6	79.6
	Yes	1307	20.4	20.4	100.0
	Total	6400	100.0	100.0	

Owns TV

		Frequency	Percent	Valid Percent	Cumulative Percent
Valid	No	63	1.0	1.0	1.0
	Yes	6337	99.0	99.0	100.0
	Total	6400	100.0	100.0	

The frequency tables are displayed in the Viewer window. The frequency tables reveal that only 20.4% of the people own PDAs, but almost everybody owns a TV (99.0%). These might not be interesting revelations, although it might be interesting to find out more about the small group of people who do not own televisions.

Charts for Categorical Data

You can graphically display the information in a frequency table with a bar chart or pie chart.

▶ Open the Frequencies dialog box again. (The two variables should still be selected.)

You can use the Dialog Recall button on the toolbar to quickly return to recently used procedures.

Figure 6-3
Dialog Recall button

▶ Click Charts.

▶ Select Bar charts and then click Continue.

Figure 6-4
Frequencies Charts dialog box

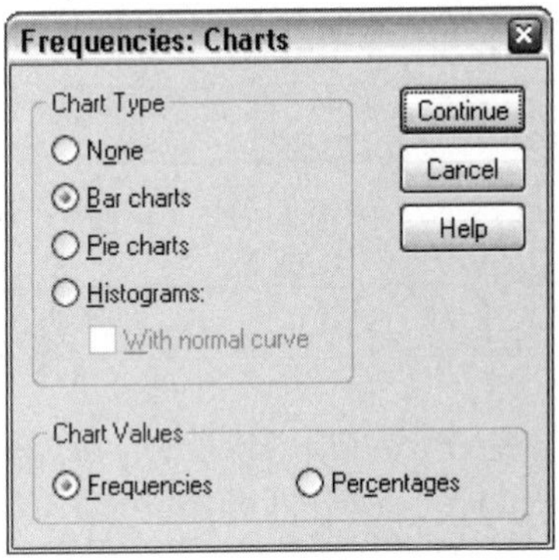

▶ Click OK in the main dialog box to run the procedure.

Figure 6-5
Bar chart

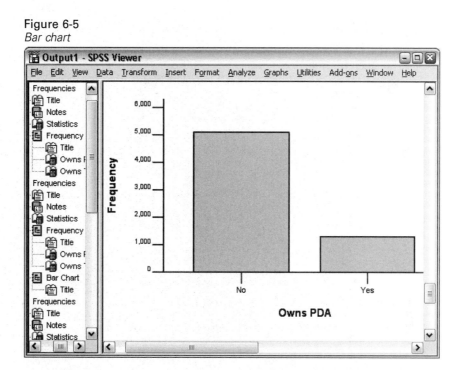

In addition to the frequency tables, the same information is now displayed in the form of bar charts, making it easy to see that most people do not own PDAs but almost everyone owns a TV.

Summary Measures for Scale Variables

There are many summary measures available for scale variables, including:

- **Measures of central tendency.** The most common measures of central tendency are the **mean** (arithmetic average) and **median** (value at which half the cases fall above and below).

- **Measures of dispersion.** Statistics that measure the amount of variation or spread in the data include the standard deviation, minimum, and maximum.

▶ Open the Frequencies dialog box again.

▶ Click Reset to clear any previous settings.

▶ Select *Household income in thousands [income]* and move it into the Variable(s) list.

Figure 6-6
Scale variable selected for analysis

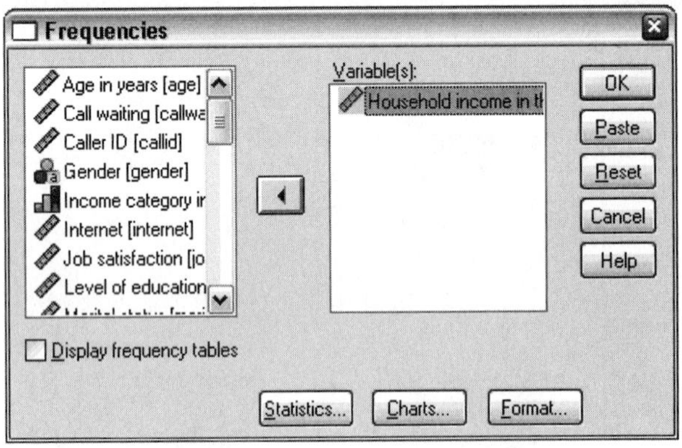

▶ Click Statistics.

▶ Select Mean, Median, Std. deviation, Minimum, and Maximum.

Figure 6-7
Frequencies Statistics dialog box

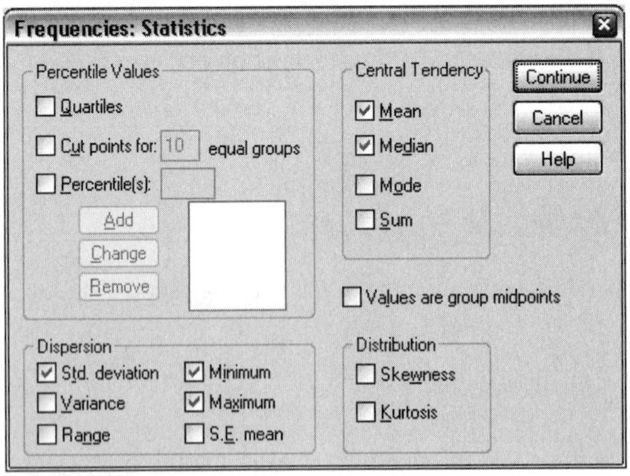

▶ Click Continue.

▶ Deselect Display frequency tables in the main dialog box. (Frequency tables are usually not useful for scale variables since there may be almost as many distinct values as there are cases in the data file.)

▶ Click OK to run the procedure.

The Frequencies Statistics table is displayed in the Viewer window.

Figure 6-8
Frequencies Statistics table

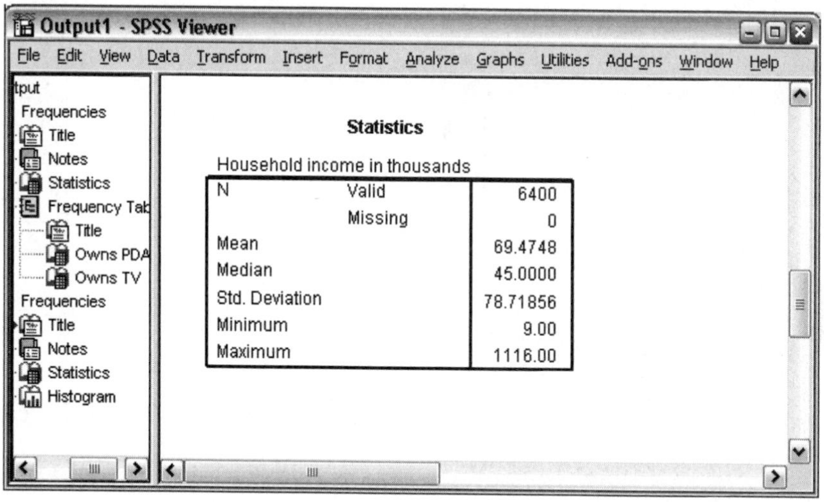

In this example, there is a large difference between the mean and the median. The mean is almost 25,000 greater than the median, indicating that the values are not normally distributed. You can visually check the distribution with a histogram.

Histograms for Scale Variables

▶ Open the Frequencies dialog box again.

▶ Click Charts.

▶ Select Histograms and With normal curve.

Figure 6-9
Frequencies Charts dialog box

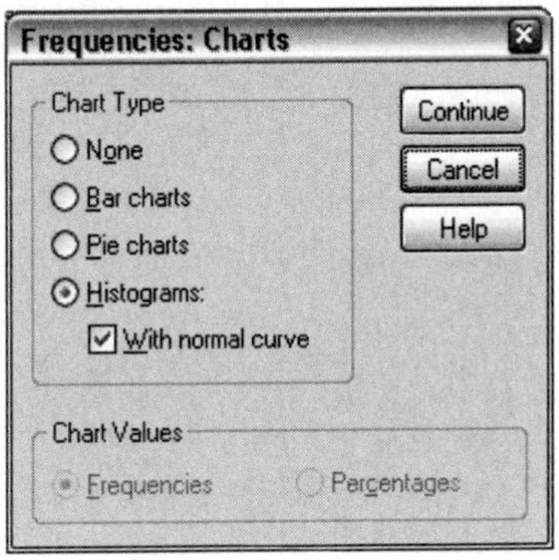

▶ Click Continue, and then click OK in the main dialog box to run the procedure.

Figure 6-10
Histogram

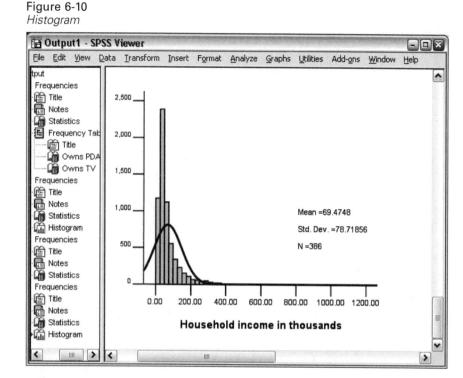

The majority of cases are clustered at the lower end of the scale, with most falling below 100,000. There are, however, a few cases in the 500,000 range and beyond (too few to even be visible without modifying the histogram). These high values for only a few cases have a significant effect on the mean but little or no effect on the median, making the median a better indicator of central tendency in this example.

Working with Output

The results from running a statistical procedure are displayed in the Viewer. The output produced can be statistical tables, charts, graphs, or text, depending on the choices you make when you run the procedure. This chapter uses the files *viewertut.spo* and *demo.sav*.

Using the Viewer

Figure 7-1
SPSS Viewer

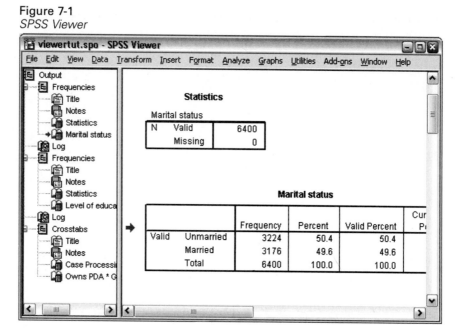

The Viewer window is divided into two panes. The **outline pane** contains an outline of all of the information stored in the Viewer. The **contents pane** contains statistical tables, charts, and text output.

Use the scroll bars to navigate through the window's contents, both vertically and horizontally. For easier navigation, click an item in the outline pane to display it in the contents pane.

If you find that there isn't enough room in the Viewer to see an entire table or that the outline view is too narrow, you can easily resize the window.

▶ Click and drag the right border of the outline pane to change its width.

An open book icon in the outline pane indicates that it is currently visible in the Viewer, although it may not currently be in the visible portion of the contents pane.

▶ To hide a table or chart, double-click its book icon in the outline pane.

The open book icon changes to a closed book icon, signifying that the information associated with it is now hidden.

▶ To redisplay the hidden output, double-click the closed book icon.

You can also hide all of the output from a particular statistical procedure or all of the output in the Viewer.

▶ Click the box with the minus sign (−) to the left of the procedure whose results you want to hide, or click the box next to the topmost item in the outline pane to hide all of the output.

Figure 7-2
Hidden output in the Viewer

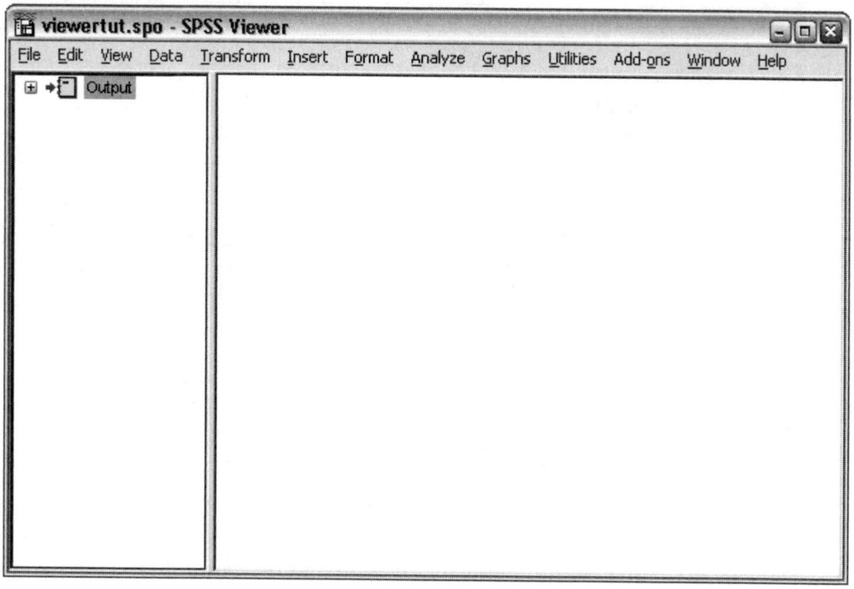

The outline collapses, visually indicating that these results are hidden.

You can also change the order in which the output is displayed.

▶ In the outline pane, click on the items that you want to move.

▶ Drag the selected items to a new location in the outline and release the mouse button.

Figure 7-3
Reordered output in the Viewer

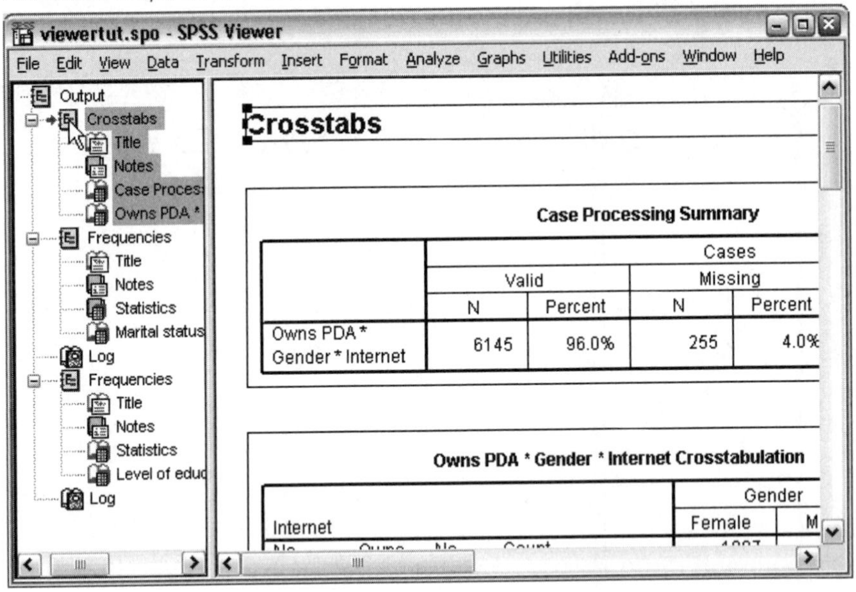

You can also move output items by clicking and dragging them in the contents pane.

Using the Pivot Table Editor

The results from most statistical procedures are displayed in **pivot tables**.

Accessing Output Definitions

Many statistical terms are displayed in the output. Definitions of these terms can be accessed directly in the Viewer.

▶ Double-click the *Owns PDA * Gender * Internet Crosstabulation* table.

▶ Right-click *Expected Count* and choose What's This? from the pop-up context menu.

The definition is displayed in a pop-up window.

Figure 7-4
Pop-up definition

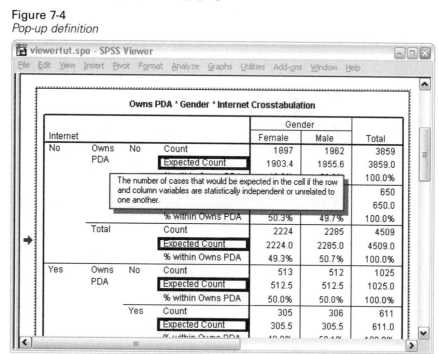

Pivoting Tables

The default tables produced may not display information as neatly or as clearly as you would like. With pivot tables, you can transpose rows and columns ("flip" the table), adjust the order of data in a table, and modify the table in many other ways. For example, you can change a short, wide table into a long, thin one by transposing rows and columns. Changing the layout of the table does not affect the results. Instead, it's a way to display your information in a different or more desirable manner.

► Double-click the *Owns PDA * Gender * Internet Crosstabulation* table.

► If the Pivoting Trays window is not visible, from the menus choose:
Pivot
    Pivoting Trays

Pivoting trays provide a way to move data between columns, rows, and layers.

Figure 7-5
Pivoting trays

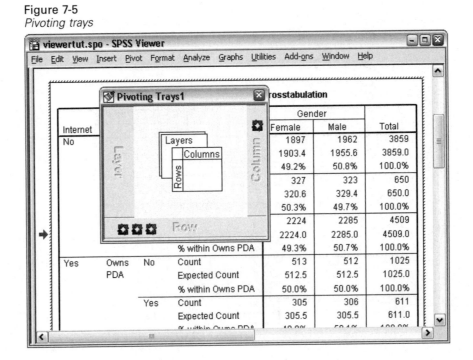

▶ Click one of the pivot icons to see what it represents. The shaded area in the table indicates what will be moved when you move the pivot icon. A pop-up label also indicates what the icon represents in the table.

Figure 7-6
Pivot icon labels

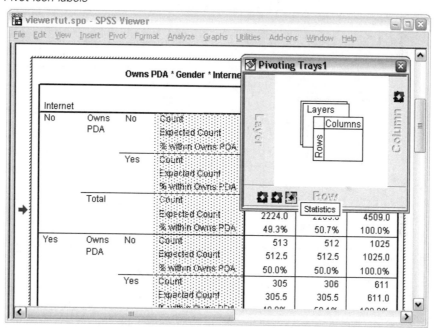

▶ Drag the Statistics pivot icon from the Row dimension to the bottom of the Column dimension. The table is immediately reconfigured to reflect your changes.

The order of the pivot icons in a dimension reflects the order of the elements in the table.

Figure 7-7
Pivoting tray icon associations

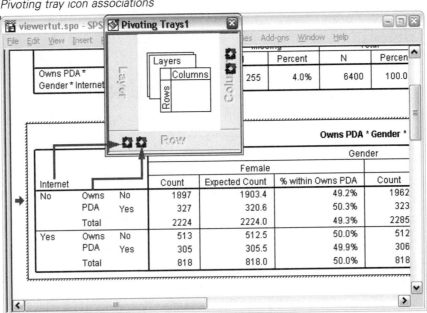

▶ Drag the *Owns PDA* icon to the left of the *Internet* icon and release the mouse button to reverse the order of these two rows.

Creating and Displaying Layers

Layers can be useful for large tables with nested categories of information. By creating layers, you simplify the look of the table, making it easier to read. Layers work best when the table has at least three variables.

▶ Double-click the *Owns PDA * Gender * Internet Crosstabulation* table.

▶ If the Pivoting Trays window is not visible, from the menus choose:
Pivot
  Pivoting Trays

▶ Drag the *Gender* pivot icon from the Column dimension to the Layer dimension.

Figure 7-8
Gender pivot icon in the Layer dimension

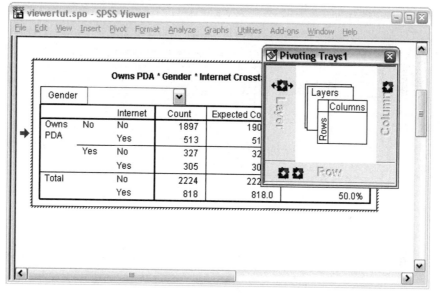

To view the different layers, you can either click the arrows on the layer pivot icon, or you can select a layer from the drop-down list in the table.

Figure 7-9
Choosing a layer

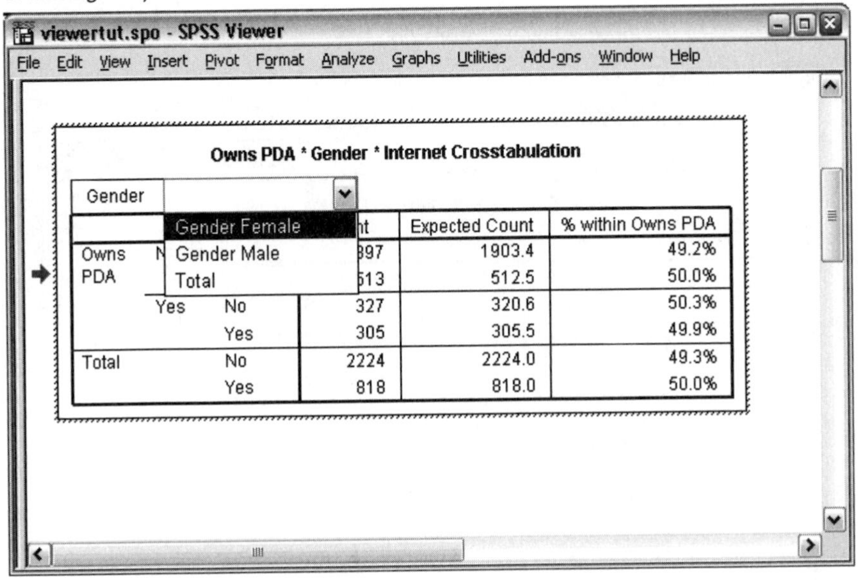

Editing Tables

Unless you've taken the time to create a custom TableLook, pivot tables are created with standard formatting. You can change the formatting of any text within a table. Formats that you can change include font name, font size, font style (bold or italic), and color.

▶ Double-click the *Level of education* table.

▶ If the Formatting toolbar is not visible, from the menus choose:
   View
     Toolbar

▶ Click the title text, *Level of education*.

▶ From the drop-down list of font sizes on the toolbar, choose 12.

- To change the color of the title text, click the Text Color button and choose a new color.

Figure 7-10
Reformatted title text in the pivot table

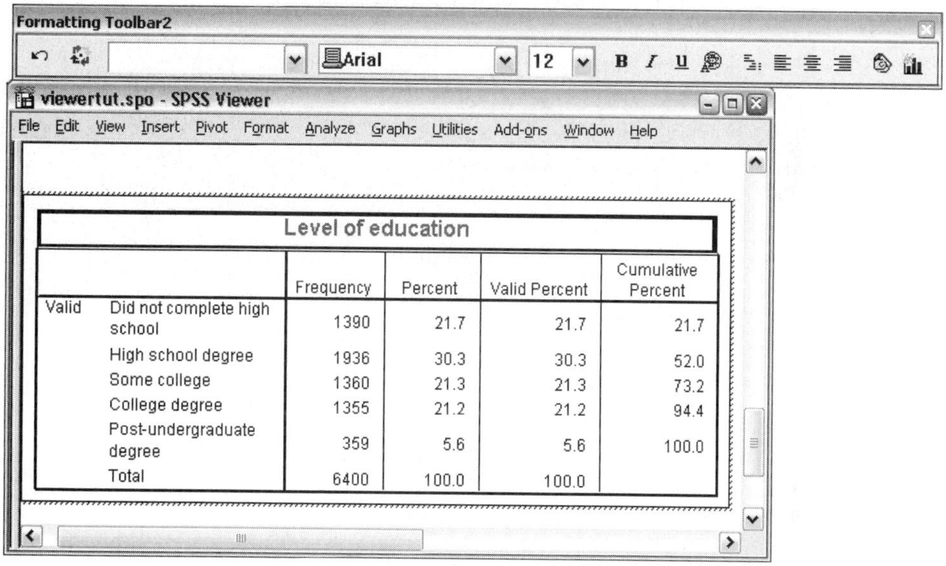

Level of education

		Frequency	Percent	Valid Percent	Cumulative Percent
Valid	Did not complete high school	1390	21.7	21.7	21.7
	High school degree	1936	30.3	30.3	52.0
	Some college	1360	21.3	21.3	73.2
	College degree	1355	21.2	21.2	94.4
	Post-undergraduate degree	359	5.6	5.6	100.0
	Total	6400	100.0	100.0	

You can also edit the contents of tables and labels. For example, you can change the title of this table.

- Double-click the title.
- Type Education Level for the new label.

Note: If you change the values in a table, totals and other statistics are not recalculated.

Hiding Rows and Columns

Some of the data displayed in a table may not be useful or it may unnecessarily complicate the table. Fortunately, you can hide entire rows and columns without losing any data.

- Double-click the *Education Level* table.
- Ctrl-Alt-click on the *Valid Percent* column label to select all of the cells in that column.

▶ Right-click the highlighted column and choose Hide Category from the pop-up context menu.

The column is now hidden but not deleted.

To redisplay the column:

▶ From the menus choose:
  View
    Show All

Rows can be hidden and displayed in the same way as columns.

Changing Data Display Formats

You can easily change the display format of data in pivot tables.

▶ Double-click the *Education Level* table.

▶ Click on the *Percent* column label to select it.

▶ From the menus choose:
  Edit
    Select
      Data Cells

▶ From the menus choose:
  Format
    Cell Properties...

▶ Type 0 in the Decimals field to hide all decimal points in this column.

Figure 7-11
Cell Properties dialog box

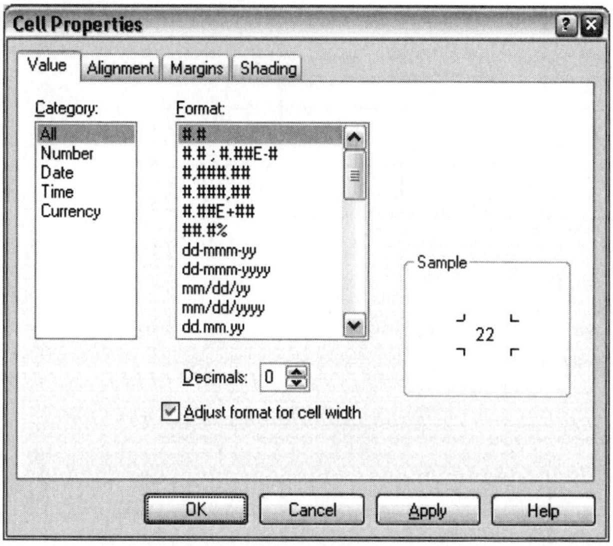

You can also change the data type and format in this dialog box.

▶ Select the type that you want from the Category list, and then select the format for that type in the Format list.

▶ Click OK to apply your changes and return to the Viewer.

Figure 7-12
Decimals hidden in Percent column

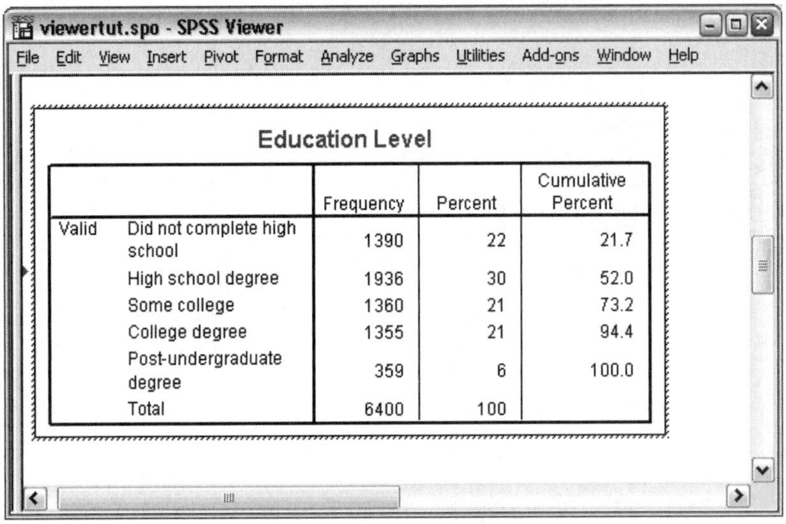

Education Level

		Frequency	Percent	Cumulative Percent
Valid	Did not complete high school	1390	22	21.7
	High school degree	1936	30	52.0
	Some college	1360	21	73.2
	College degree	1355	21	94.4
	Post-undergraduate degree	359	6	100.0
	Total	6400	100	

The decimals are now hidden in the *Percent* column.

TableLooks

The format of your tables is a critical part of providing clear, concise, and meaningful results. If your table is difficult to read, the information contained within that table may not be easily understood.

Using Predefined Formats

▶ Double-click the *Marital status* table.

▶ From the menus choose:

Format
TableLooks...

The TableLooks dialog box lists a variety of predefined styles. Select a style from the list to preview it in the Sample window on the right.

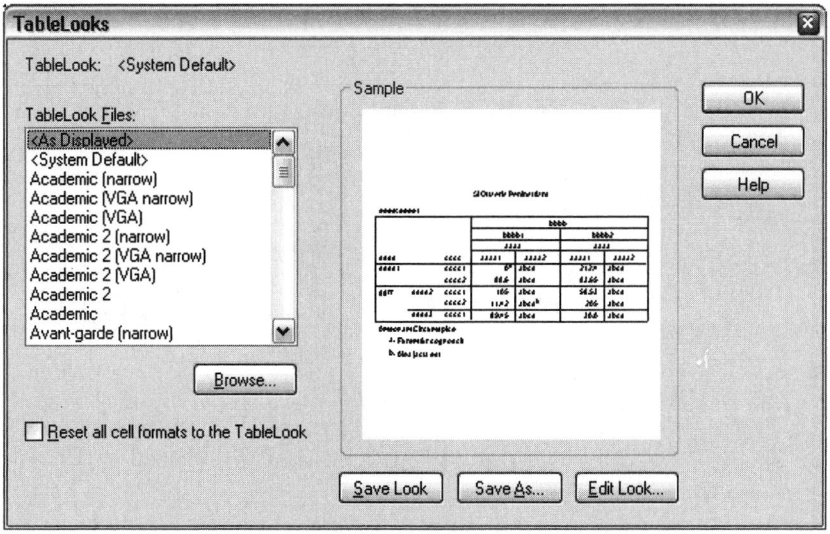

Figure 7-13
TableLooks dialog box

You can use a style as is, or you can edit an existing style to better suit your needs.

▶ To use an existing style, select one and click OK.

Customizing TableLook Styles

You can customize a format to fit your specific needs. Almost all aspects of a table can be customized, from the background color to the border styles.

▶ Double-click the *Marital status* table.

▶ From the menus choose:
Format
    TableLooks...

▶ Select the style that is closest to your desired format and click Edit Look.

▶ Click the Cell Formats tab to view the formatting options.

Figure 7-14
Table Properties dialog box

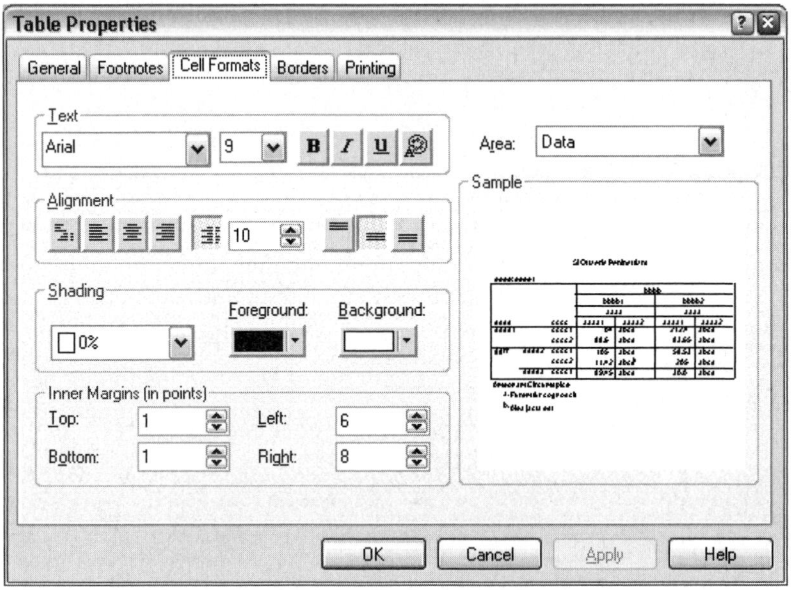

The formatting options include font name, font size, style, and color. Additional options include alignment, shading, foreground and background colors, and margin sizes.

The Sample window on the right provides a preview of how the formatting changes affect your table. Each area of the table can have different formatting styles. For example, you probably wouldn't want the title to have the same style as the data. To select a table area to edit, you can either choose the area by name in the Area drop-down list, or you can click the area that you want to change in the Sample window.

▶ Select Title from the Area drop-down list.

▶ Select a new color from the Background drop-down list.

The Sample window shows the new style.

▶ Click OK to return to the TableLooks dialog box.

You can save your new style, which allows you to apply it to future tables easily.

▶ Click Save As.

▶ Navigate to the desired target directory and enter a name for your new style in the File Name text box.

▶ Click Save.

▶ Click OK to apply your changes and return to the Viewer.

The table now contains the custom formatting that you specified.

Figure 7-15
Custom TableLook

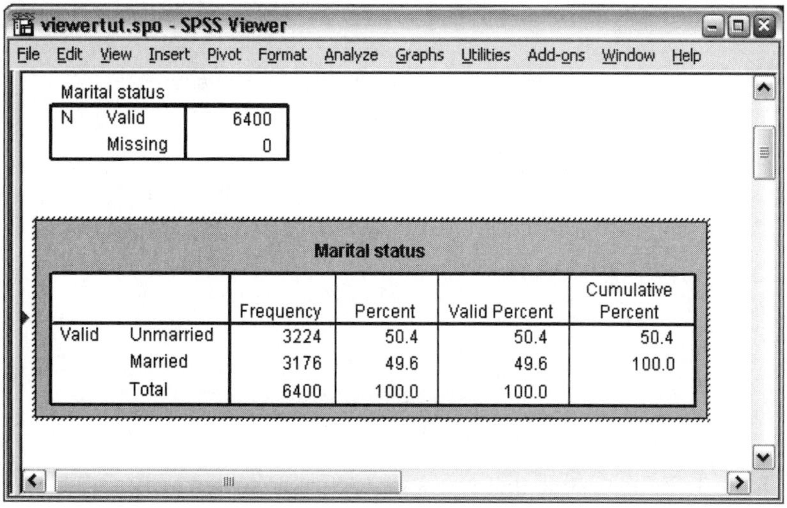

Changing the Default Table Formats

Although you can change the format of a table after it has been created, it may be more efficient to change the default TableLook so that you do not have to change the format every time you create a table.

To change the default TableLook style for your pivot tables, from the menus choose:
Edit
  Options...

▶ Click the Pivot Tables tab in the Options dialog box.

Figure 7-16
Options dialog box

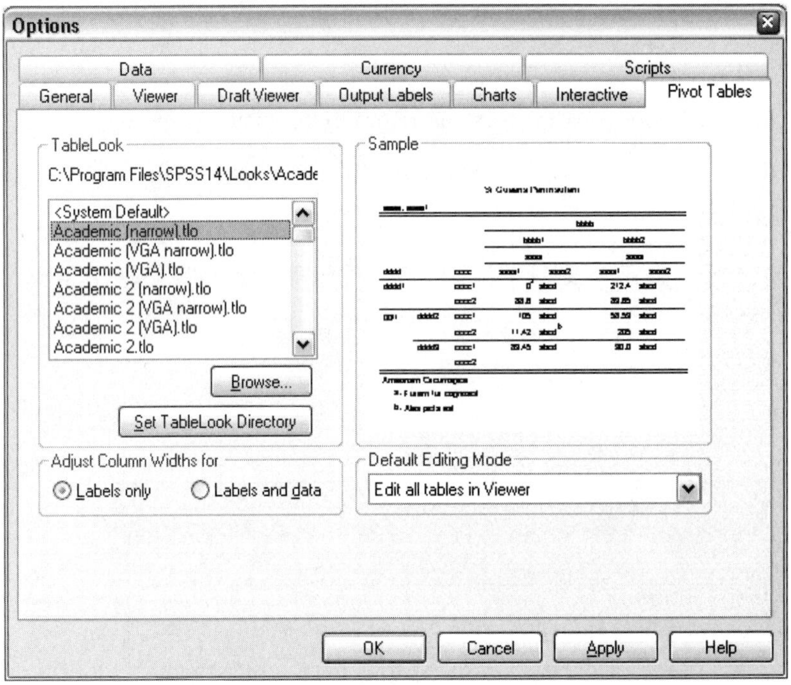

▶ Select the TableLook style that you want to use for all new tables.

The Sample window on the right shows a preview of each TableLook.

▶ Click OK to save your settings and close the dialog box.

All tables that you create after changing the default TableLook automatically conform to the new formatting rules.

Customizing the Initial Display Settings

The initial display settings include the alignment of objects in the Viewer, whether objects are shown or hidden by default, and the width of the Viewer window. To change these settings:

▶ From the menus choose:
Edit
  Options...

▶ Click the Viewer tab.

Figure 7-17
Viewer options

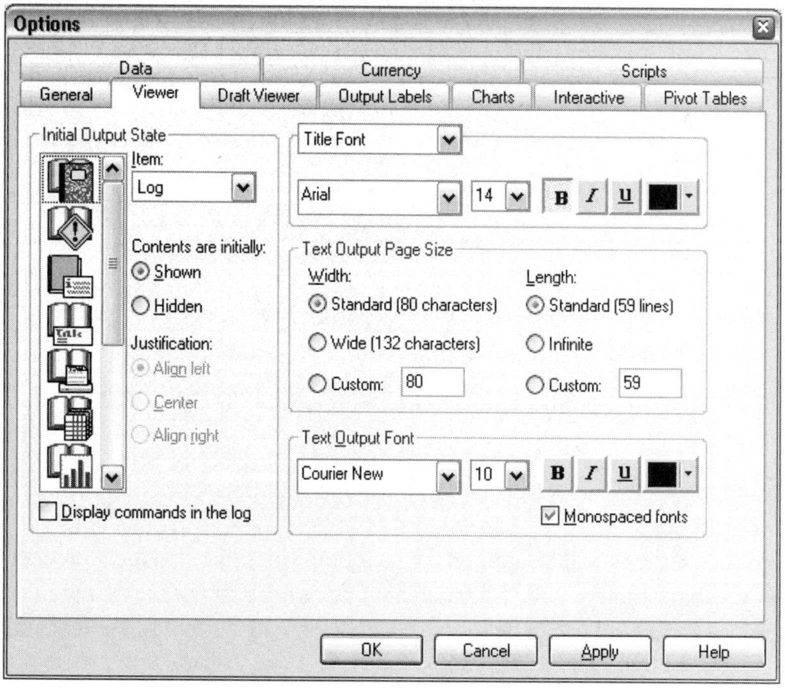

The settings are applied on an object-by-object basis. For example, you can customize the way charts are displayed without making any changes to the way tables are displayed. Simply select the object that you want to customize, and make the desired changes.

▶ Click the Title icon to display its settings.

▶ Click Center to display all titles in the (horizontal) center of the Viewer.

You can also hide elements, such as the log and warning messages, that tend to clutter your output. Double-clicking on an icon automatically changes that object's display property.

▶ Double-click the Warnings icon to hide warning messages in the output.

▶ Click OK to save your changes and close the dialog box.

Figure 7-18
Centered title in the Marital status table

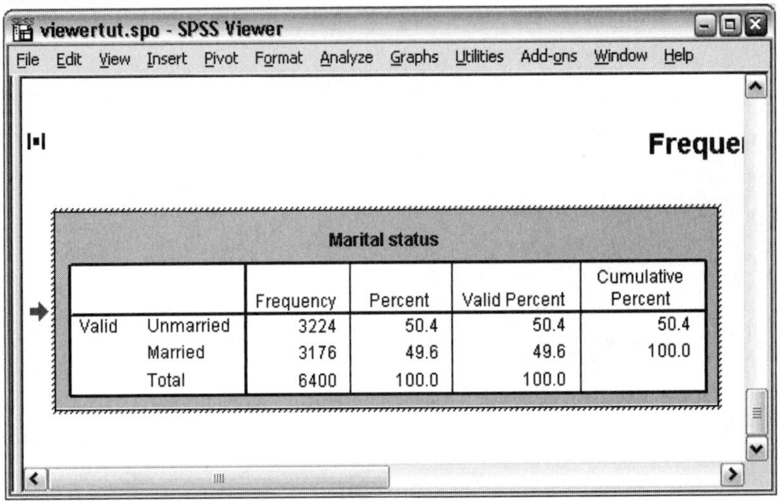

	viewertut.spo - SPSS Viewer				
File	Edit	View	Insert	Pivot	Format	Analyze	Graphs	Utilities	Add-ons	Window	Help

Freque

Marital status					
		Frequency	Percent	Valid Percent	Cumulative Percent
Valid	Unmarried	3224	50.4	50.4	50.4
	Married	3176	49.6	49.6	100.0
	Total	6400	100.0	100.0	

Your new settings will be applied the next time you run a statistical procedure. All items that you have hidden will still be created, but they will not be visible in the contents pane. Centered items can be identified by the small symbol to the left of the item.

Displaying Variable and Value Labels

In most cases, displaying the labels for variables and values is more effective than displaying the variable name or the actual data value. There may be cases, however, when you want to display both the names and the labels.

▶ From the menus choose:
Edit
  Options...

▶ Click the Output Labels tab.

Figure 7-19
Output Labels options

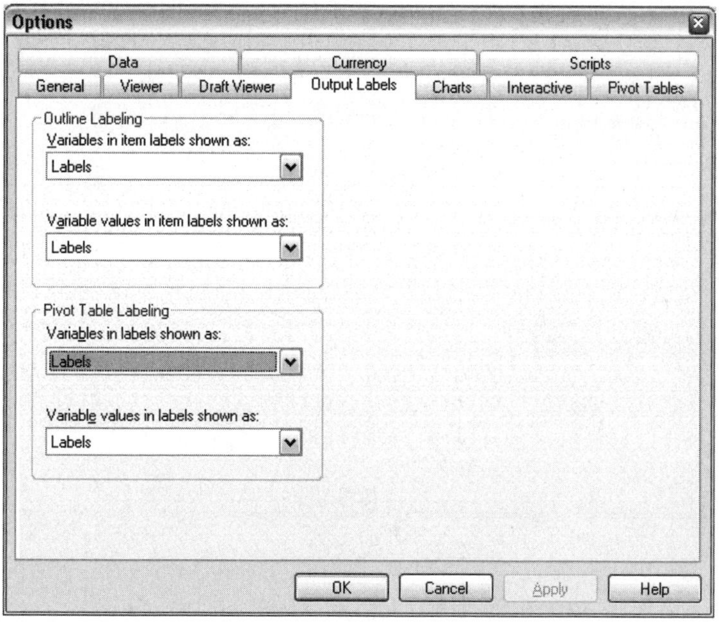

You can specify different settings for the outline and contents panes. For example, to show labels in the outline and variable names and data values in the contents:

▶ In the Pivot Table Labeling group, select Names from the Variables in Labels drop-down list to show variable names instead of labels.

▶ Then, select Values from the Variable Values in Labels drop-down list to show data values instead of labels.

Figure 7-20
Pivot Table Labeling settings

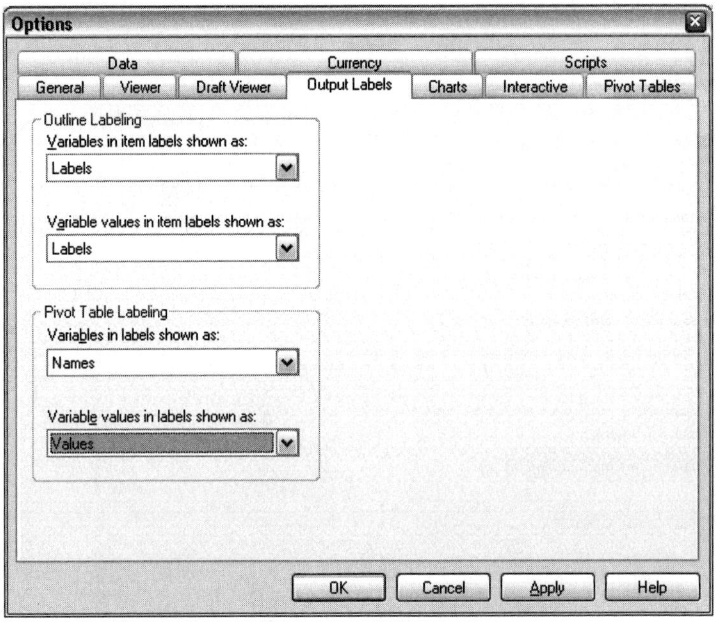

The new settings are applied the next time you run a statistical procedure.

Figure 7-21
Variable names and values displayed

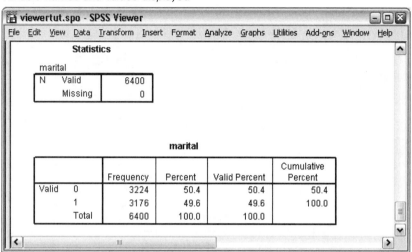

Using Results in Other Applications

Your results can be used in many applications. For example, you may want to include a chart or graph in a presentation or report. Applications such as Microsoft PowerPoint or Word can display your results as plain text, rich text, or as a metafile, which is a graphical representation of the output.

The following examples are specific to Microsoft Word, but they may work similarly in other word processing applications.

Pasting Results as Word Tables

You can paste pivot tables into Word as native Word tables. All table attributes, such as font sizes and colors, are retained. Because the table is pasted in the Word table format, you can edit it in Word just like any other table.

▶ Click the *Marital status* table in the Viewer.

▶ From the menus choose:
Edit
  Copy

▶ Open your word processing application.

▶ From the word processor's menus choose:
Edit
  Paste Special...

▶ Select Formatted Text (RTF) in the Paste Special dialog box.

Figure 7-22
Paste Special dialog box

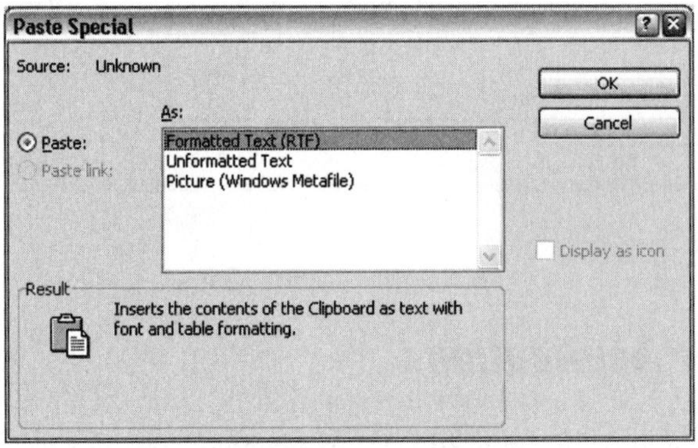

▶ Click OK to paste your results into the current document.

Figure 7-23
Pivot table displayed in Word

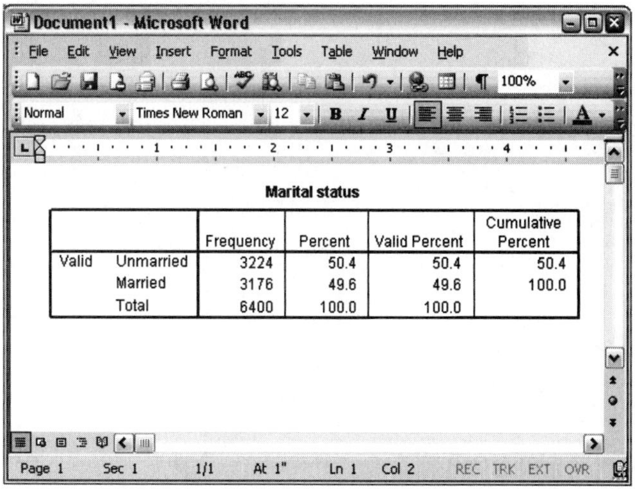

The table is now displayed in your document. You can apply custom formatting, edit the data, and resize the table to fit your needs.

Pasting Results as Metafiles

Pasting your results as metafiles maintains the original look of the output, but the pasted output becomes a vector graphic image in the target document.

▶ Click the *Marital status* table in the Viewer.

▶ From the menus choose:
Edit
 Copy

▶ Open your word processing application.

▶ From the word processor's menus choose:
Edit
 Paste Special...

▶ Select Picture in the Paste Special dialog box. (In some applications, the choice may be "Metafile" instead of "Picture.")

Figure 7-24
Paste Special dialog box

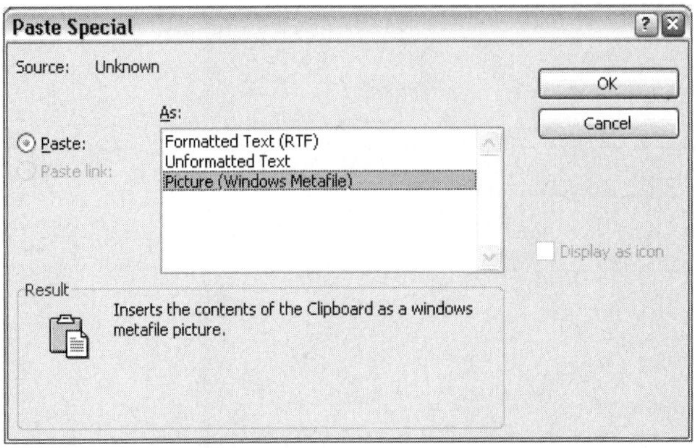

▶ Click OK to paste your results into the current document.

Figure 7-25
Metafile displayed in Word

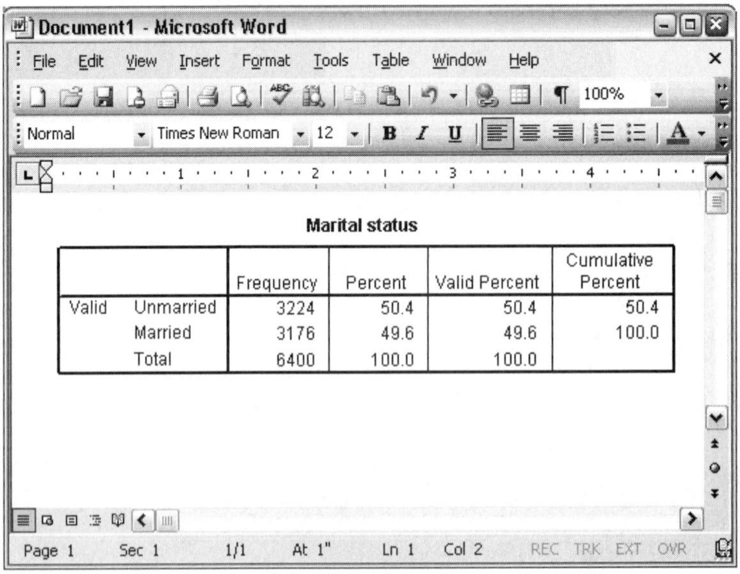

Marital status

		Frequency	Percent	Valid Percent	Cumulative Percent
Valid	Unmarried	3224	50.4	50.4	50.4
	Married	3176	49.6	49.6	100.0
	Total	6400	100.0	100.0	

The metafile is now embedded in your document. This image is a snapshot of the *Marital status* table. Only the visible portions of the table are copied. Information in hidden categories or layers is not included in the metafile.

Pasting Results as Text

Pivot tables can be copied to other applications as plain text. Formatting styles are not retained in this method, but you gain the ability to edit the table data after you paste it into the target application.

▶ Click the *Marital status* table in the Viewer.

▶ From the menus choose:
Edit
  Copy

▶ Open your word processing application.

▶ From the word processor's menus choose:
Edit
  Paste Special...

▶ Select Unformatted Text in the Paste Special dialog box.

Figure 7-26
Paste Special dialog box

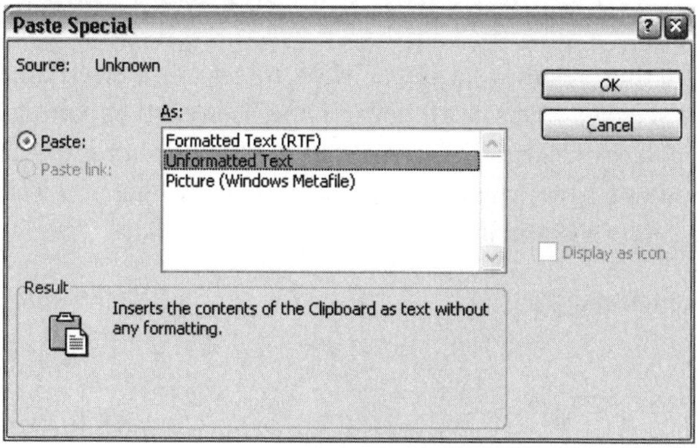

▶ Click OK to paste your results into the current document.

Figure 7-27
Pivot table displayed in Word

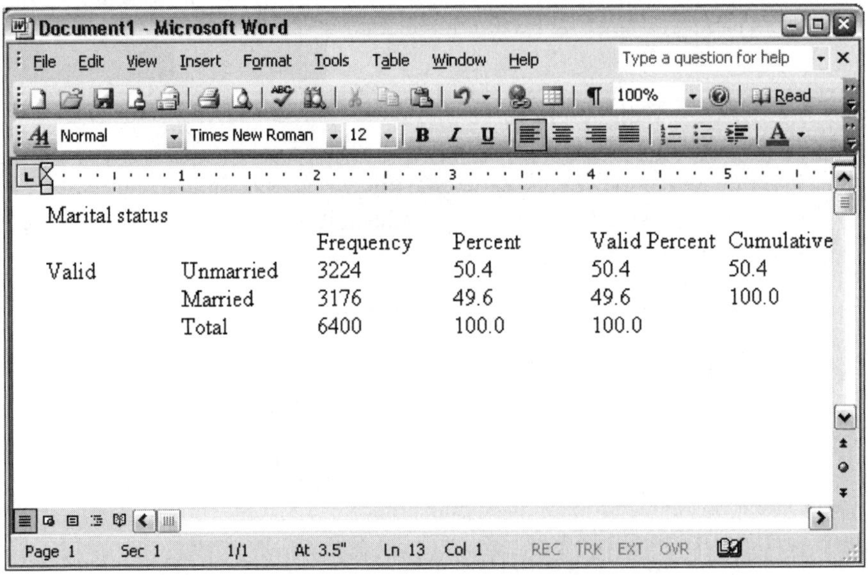

Each column of the table is separated by tabs. You can change the column widths by adjusting the tab stops in your word processing application.

Exporting Results to Microsoft Word, PowerPoint, and Excel Files

SPSS allows you to export results to a single Microsoft Word, PowerPoint, or Excel file. You can export selected items or all items in the Viewer. If exporting to Word or PowerPoint, you can export charts. This topic uses the files *msouttut.spo* and *demo.sav*.

In the Viewer's outline pane, you can select specific items that you want to export. You do not have to select specific items.

▶ From the Viewer menus choose:
File
  Export...

There are several options for exporting the results.

Figure 7-28
Export Output dialog box

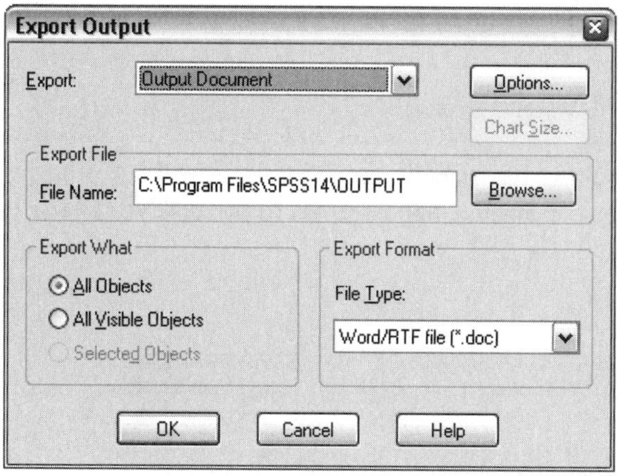

First, you can select which type of output file you want to create. For Word and PowerPoint, you can create a file that contains charts (Output Document) or one that does not contain charts (Output Document (No Charts)). Charts are embedded in Word documents as Windows metafiles. For Excel, you can create only documents that do not contain charts (Output Document (No Charts)).

▶ Select Output Document from the Export drop-down list.

Note: Export to PowerPoint is not available with the Student Version.

You can save the exported Word, PowerPoint, or Excel file to any location and assign it any name that Windows allows. The default file name is *OUTPUT*, and the default path is the installation location or, if you opened or saved an output file, the location of that file. You can change the default by entering a new path or by clicking Browse to locate a destination. You do not have to specify a file extension. The export functionality adds the appropriate extension.

Note the default file path in the File Name text box. You will need to know where the file is to open it.

Instead of exporting all objects in the Viewer, you can choose to export only visible objects (open books in the outline pane) or those that you selected in the outline pane. If you did not select any items in the outline pane, you do not have the option to export selected objects.

▶ Select All Objects in the Export What group.

Finally, you select the file format.

▶ Select Word/RTF file (*.doc) from the File Type drop-down list.

▶ Click OK to generate the Word file.

When you open the resulting file in Word, you can see how the results are exported. Notes, which are not visible objects, appear in Word because you chose to export all objects.

Figure 7-29
Output.doc in Word

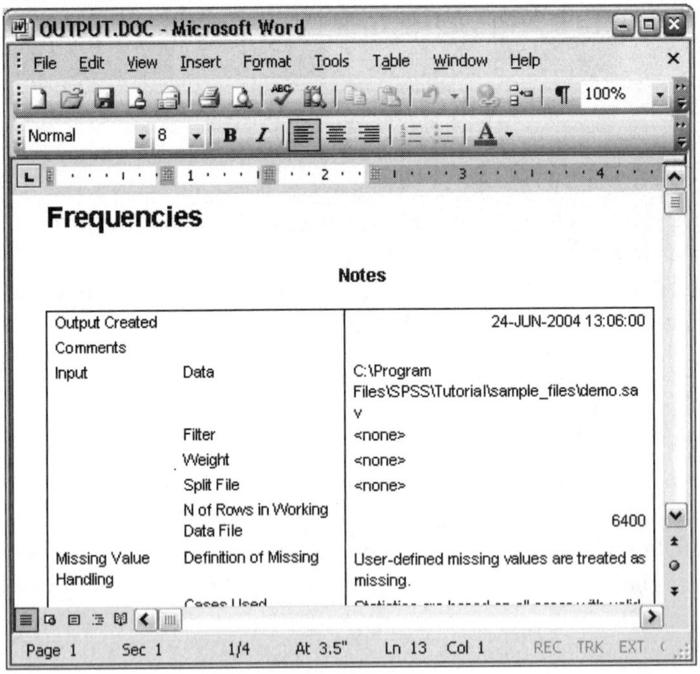

Pivot tables become Word tables, with all of the formatting of the original pivot table retained, including fonts, colors, borders, and so on.

Figure 7-30
Pivot tables in Word

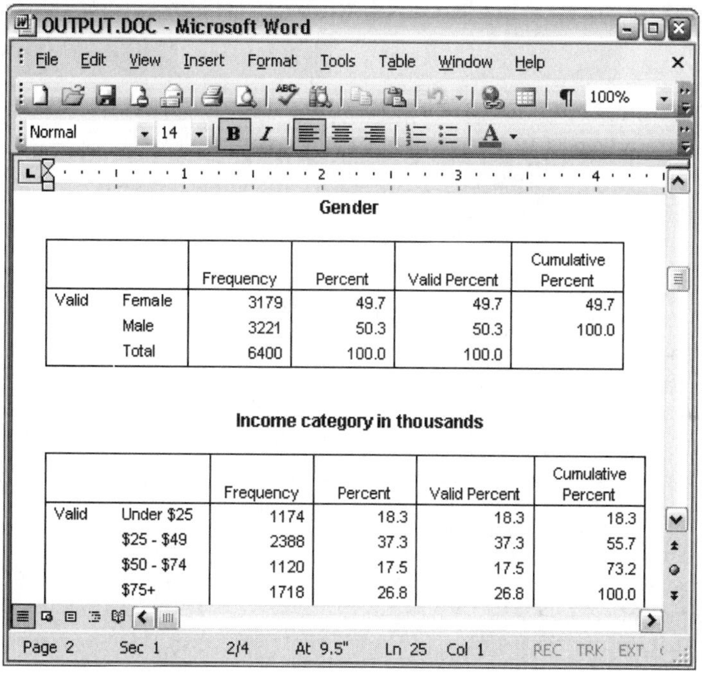

OUTPUT.DOC - Microsoft Word

File   Edit   View   Insert   Format   Tools   Table   Window   Help

Normal   14   **B**   *I*

Gender

		Frequency	Percent	Valid Percent	Cumulative Percent
Valid	Female	3179	49.7	49.7	49.7
	Male	3221	50.3	50.3	100.0
	Total	6400	100.0	100.0	

Income category in thousands

		Frequency	Percent	Valid Percent	Cumulative Percent
Valid	Under $25	1174	18.3	18.3	18.3
	$25 - $49	2388	37.3	37.3	55.7
	$50 - $74	1120	17.5	17.5	73.2
	$75+	1718	26.8	26.8	100.0

Page 2   Sec 1   2/4   At 9.5"   Ln 25   Col 1   REC   TRK   EXT

Charts become embedded Windows metafiles.

Figure 7-31
Charts in Word

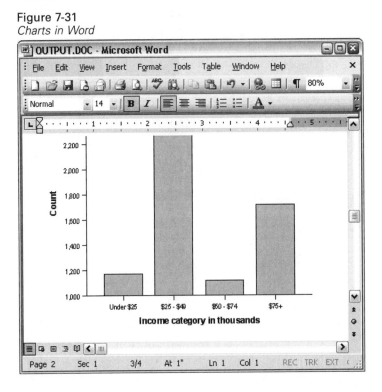

Text output is displayed in a fixed-pitch font.

Figure 7-32
Text output in Word

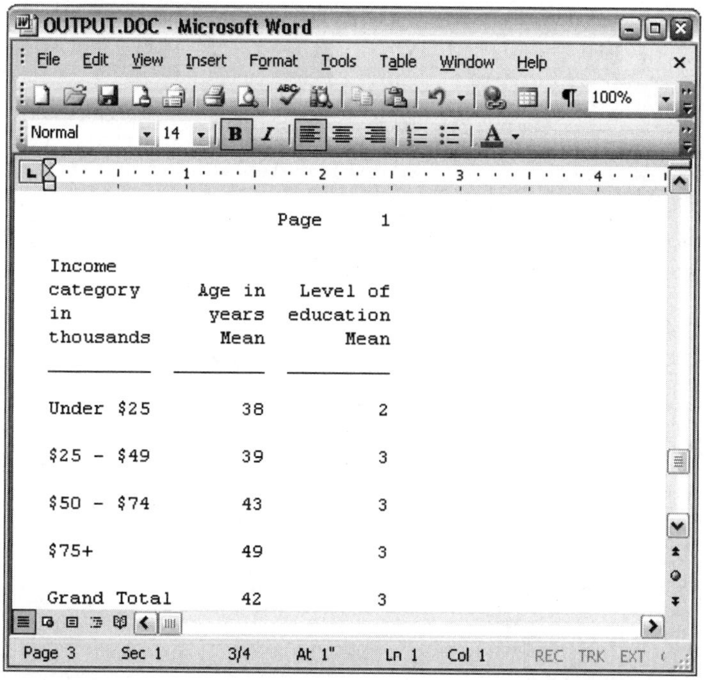

If you export to a PowerPoint file, each exported item is placed on a separate slide. Pivot tables exported to PowerPoint become Word tables, with all of the formatting of the original pivot table, including fonts, colors, borders, and so on.

Figure 7-33
Pivot tables in PowerPoint

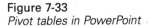

Charts selected for export to PowerPoint are embedded in the PowerPoint file.

Figure 7-34
Charts in PowerPoint

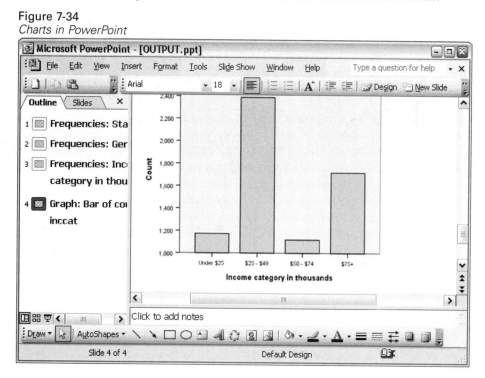

Note: Export to PowerPoint is not available with the Student Version.

If you export to an Excel file, results are exported differently.

Figure 7-35
Output.xls in Excel

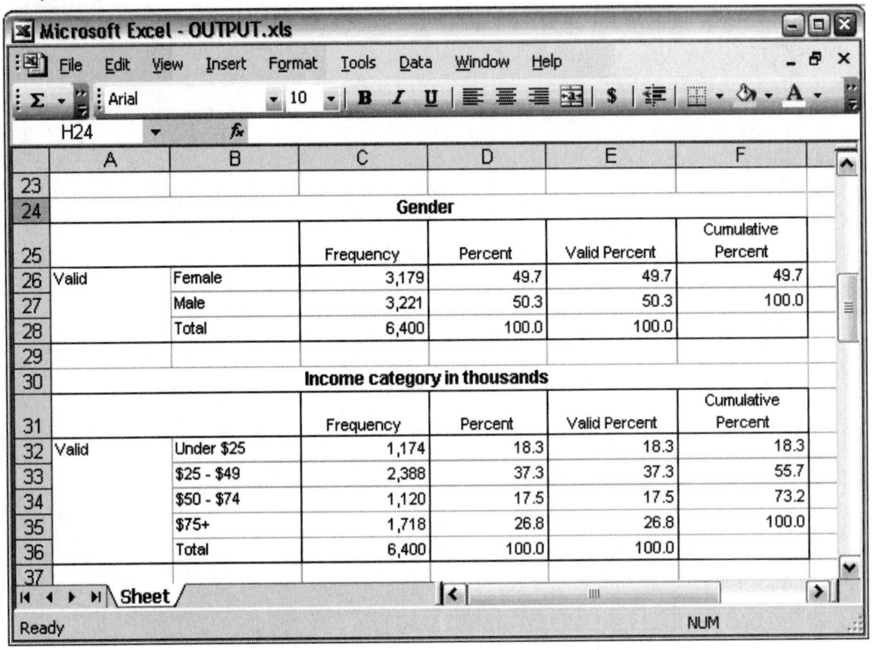

	A	B	C	D	E	F
23						
24			**Gender**			
25			Frequency	Percent	Valid Percent	Cumulative Percent
26	Valid	Female	3,179	49.7	49.7	49.7
27		Male	3,221	50.3	50.3	100.0
28		Total	6,400	100.0	100.0	
29						
30			**Income category in thousands**			
31			Frequency	Percent	Valid Percent	Cumulative Percent
32	Valid	Under $25	1,174	18.3	18.3	18.3
33		$25 - $49	2,388	37.3	37.3	55.7
34		$50 - $74	1,120	17.5	17.5	73.2
35		$75+	1,718	26.8	26.8	100.0
36		Total	6,400	100.0	100.0	
37						

Pivot table rows, columns, and cells become Excel rows, columns, and cells.

Each line in the text output is a row in the Excel file, with the entire contents of the line contained in a single cell. Charts are not exported at all.

Figure 7-36
Text output in Excel

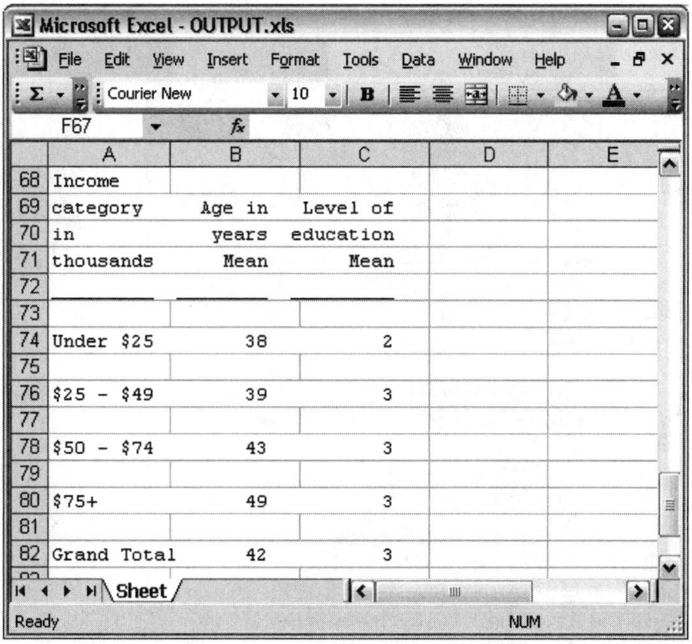

Exporting Results to HTML

You can also export results to HTML (hypertext markup language). When saving as HTML, all non-graphic output from SPSS is exported into a single HTML file.

Figure 7-37
Output.htm in Web browser

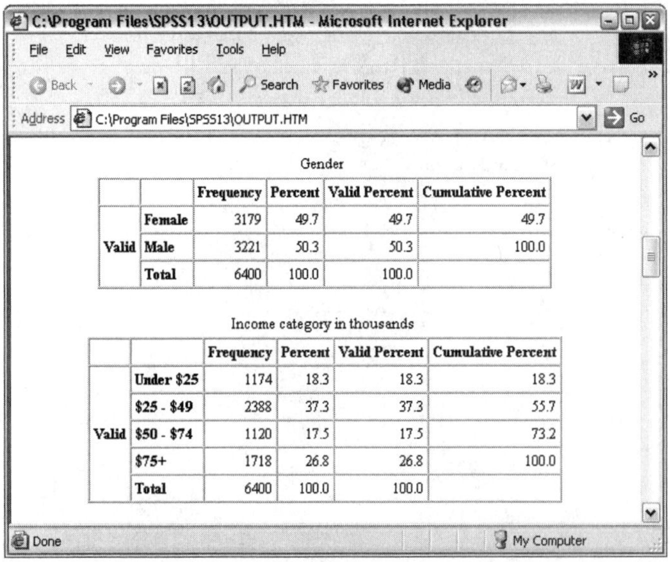

Gender

		Frequency	Percent	Valid Percent	Cumulative Percent
Valid	Female	3179	49.7	49.7	49.7
	Male	3221	50.3	50.3	100.0
	Total	6400	100.0	100.0	

Income category in thousands

		Frequency	Percent	Valid Percent	Cumulative Percent
Valid	Under $25	1174	18.3	18.3	18.3
	$25 - $49	2388	37.3	37.3	55.7
	$50 - $74	1120	17.5	17.5	73.2
	$75+	1718	26.8	26.8	100.0
	Total	6400	100.0	100.0	

When you export to HTML, charts can be exported as well, but not to a single file.

Each chart will be saved as a file in a format that you specify, and references to these graphics files will be placed in the HTML created by SPSS. There is also an option to export all charts (or selected charts) in separate graphics files.

Figure 7-38
References to graphics in HTML output

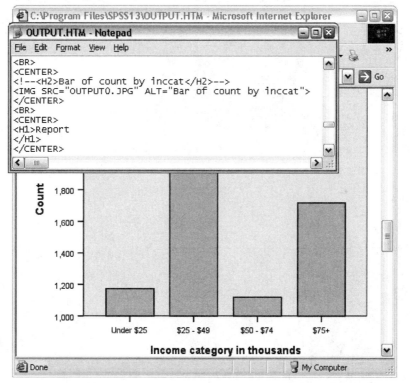

Chapter

Creating and Editing Charts

You can create and edit a wide variety of chart types in SPSS. In these examples, we will create and edit three commonly used types of charts:

- Simple bar chart
- Pie chart
- Scatterplot with groups

Chart Creation Basics

To demonstrate the basics of chart creation, we will create a bar chart of mean income for different levels of job satisfaction. This example uses the data file *demo.sav*.

▶ From the menus choose:
Graphs
  Chart Builder...

The Chart Builder dialog box is an interactive window that allows you to preview how a chart will look while you build it.

Figure 8-1
Chart Builder dialog box

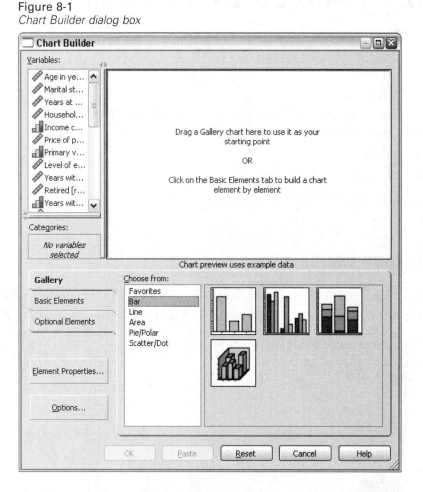

In this example, we will create a simple, two-dimensional bar chart. The resulting chart will show a bar for each group (category) in a single categorical variable. The height of each bar will be determined by the result of a statistical calculation.

Using the Chart Builder Gallery

▶ Click the Gallery tab if it is not selected.

The Gallery includes many different predefined charts, which are organized by chart type. The Basic Elements tab also provides basic elements (such as axes and data elements) for creating charts from scratch, but it's easier to use the Gallery.

▶ Click Bar if it is not selected.

Icons representing the available bar charts in the Gallery appear on the right side of the dialog box. The pictures should provide enough information to identify the specific chart type. If you need more information, you can also display a ToolTip description of the chart by pausing your cursor over an icon.

▶ Drag the icon for the simple bar chart onto the "canvas," which is the large area above the Gallery. The Chart Builder displays a preview of the chart on the canvas. Note that the data used to draw the chart are not your actual data. They are example data.

Figure 8-2
Bar chart on Chart Builder canvas

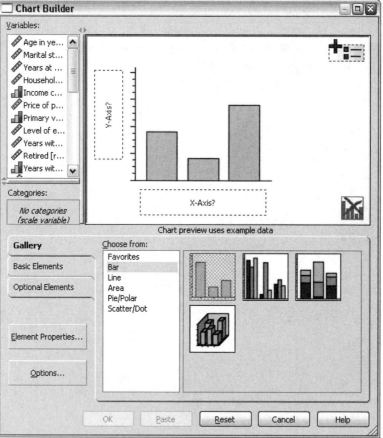

Defining Variables and Statistics

Although there is a chart on the canvas, it is not complete because there are no variables or statistics to control how tall the bars are and to specify which variable category corresponds to each bar. You can't have a chart without any variables and statistics. You add variables by dragging them from the Variables list, which is located to the left of the canvas.

When you drag the variables, the targets are "drop zones" on the canvas. Some drop zones require a variable while others do not. The drop zone for the x axis is required. The variable in this drop zone controls where the bars appear on the x axis. Depending on the type of chart you are creating, you may also need a variable in the y axis drop zone. For example, when you want to display a summary statistic of another variable (such as mean of salary), you need a variable in the y axis drop zone. Scatterplots also require a variable in the y axis. In that case, the drop zone identifies the dependent variable.

You are going to create a chart that shows bars for the mean income of each job satisfaction category, so both drop zones are needed. There will be a categorical variable on the x axis and a scale variable on the y axis for calculating the mean.

A variable's measurement level is important in the Chart Builder. You are going to use the *Job satisfaction* variable on the x axis. However, the icon (which looks like a ruler) next to the variable indicates that its measurement level is defined as scale. To create the correct chart, you must use a categorical measurement level. Instead of going back and changing the measurement level in the Variable View, you can change the measurement level temporarily in the Chart Builder.

▶ Right-click *Job satisfaction* in the Variables list and choose Ordinal. Ordinal is an appropriate measurement level because the categories in *Job satisfaction* can be ranked by level of satisfaction. Note that the icon changes after you change the measurement level.

▶ Now drag *Job satisfaction* from the Variables list to the *x* axis drop zone.

Figure 8-3
Job satisfaction in x axis drop zone

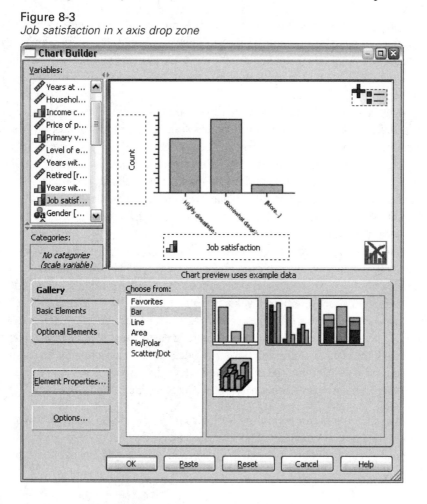

The *y* axis drop zone defaults to the *Count* statistic. If you want to use another statistic (such as percentage or mean), you can easily change it. You will not use either of these statistics in this example, but we will review the process in case you need to change this statistic at another time.

▶ Click Element Properties to display the Element Properties window.

Figure 8-4
Element Properties window

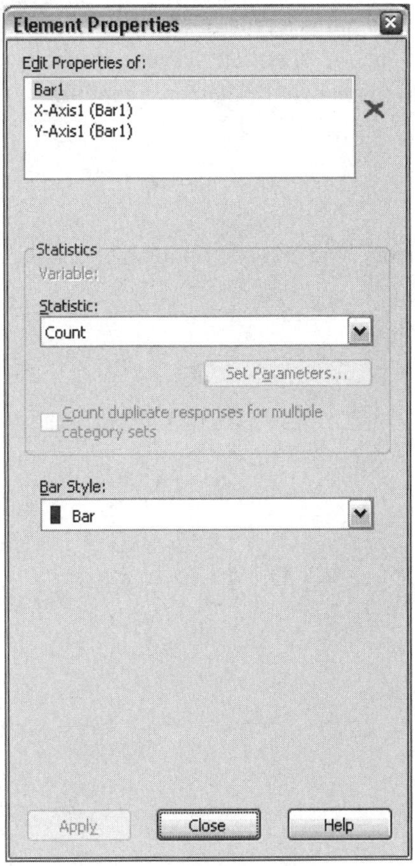

The Element Properties window allows you to change the properties of the various chart elements. These elements include the data elements (such as the bars in the bar chart) and the axes on the chart. Select one of the elements in the Edit Properties of list to change the properties associated with that element. Also note the red X located to the right of the list. This button deletes a data element from the canvas. Because Bar1 is selected, the properties shown apply to data elements, specifically the bar data element.

The Statistic drop-down list shows the specific statistics that are available. The same statistics are usually available for every chart type. Be aware that some statistics require that the y axis drop zone contains a variable.

▶ Return to the Chart Builder dialog box and drag *Household income in thousands* from the Variables list to the *y* axis drop zone. Because the variable on the *y* axis is scalar and the *x* axis variable is categorical (ordinal is a type of categorical measurement level), the *y* axis drop zone defaults to the *Mean* statistic. These are the variables and statistics you want, so there is no need to change the element properties.

Adding Optional Elements

In addition to the required chart elements, you can add optional elements, such as titles and footnotes.

▶ Click the Optional Elements tab.

Like the pictures in the Gallery, the available optional elements are items that you can drag onto the canvas.

Figure 8-5
Optional elements

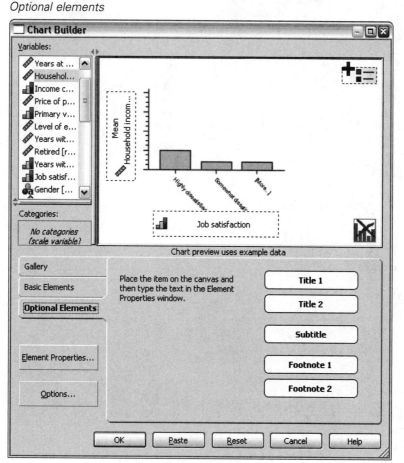

▶ Drag Title 1 to any location on the canvas.

Figure 8-6
Title 1 displayed on canvas

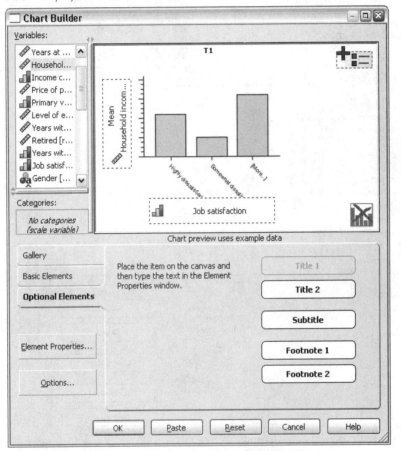

The title appears on the canvas with the label T1.

▶ To specify the title's text, click Element Properties.

▶ In the Element Properties window, select Title 1 in the Edit Properties of list.

▶ In the Content text box, type Income by Job Satisfaction. This is the text that the title will display.

▶ Click Apply to save the text. Although the text is not displayed in the Chart Builder, it will appear when you generate the chart.

Creating the Chart

▶ Click OK to create the bar chart.

Figure 8-7
Bar chart

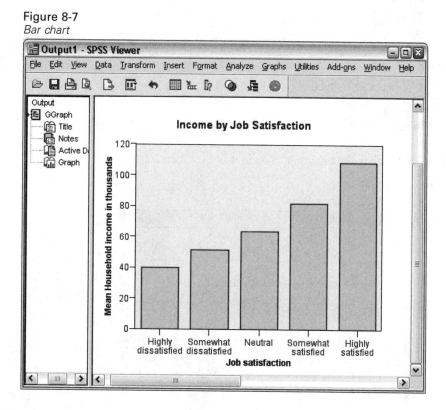

The bar chart reveals that respondents who are more satisfied with their jobs tend to have higher household incomes.

Chart Editing Basics

You can edit charts in a variety of ways. For the sample bar chart that you created, you will:

- Change colors.
- Format numbers in tick labels.
- Edit text.
- Display data value labels.
- Use chart templates.

To edit the chart, open it in the Chart Editor.

▶ Double-click the bar chart to open it in the Chart Editor.

Figure 8-8
Bar chart in the Chart Editor

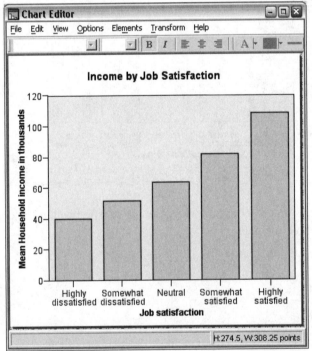

Selecting Chart Elements

To edit a chart element, you first select it.

▶ Click any one of the bars. The rectangles around the bars indicate that they are selected.

There are general rules for selecting elements in simple charts:

■ When no data elements are selected, click any data element to select all data elements.

■ When all data elements are selected, click a data element to select only that data element. You can select a different data element by clicking it. To select multiple data elements, click each element while pressing the Ctrl key.

Note: The behavior is slightly different for grouped charts. Grouped charts are discussed in the scatterplot example. For more information, see "Selecting Elements in Grouped Charts" on p. 181.

▶ To deselect all elements, press the Esc key.

▶ Click any bar to select all of the bars again.

Using the Properties Window

▶ From the Chart Editor menus choose:
Edit
  Properties

This opens the Properties window, showing the tabs that apply to the bars you selected. These tabs change depending on what chart element you select in the Chart Editor. For example, if you had selected a text frame instead of bars, different tabs would appear in the Properties window. You will use these tabs to do most chart editing.

Figure 8-9
Properties window

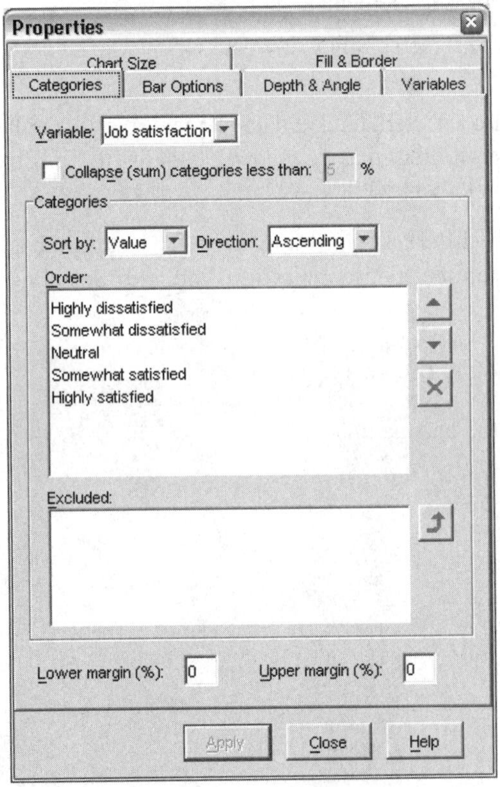

Changing Bar Colors

First, you will change the color of the bars. You specify color attributes of data elements (excluding lines and markers) on the Fill & Border tab.

▶ Click the Fill & Border tab.

▶ Click the swatch next to Fill to indicate that you want to change the fill color of the bars. The numbers below the swatch specify the red, green, and blue settings for the current color.

▶ Click the light blue color, which is second from the left in the second row from the bottom.

Figure 8-10
Fill & Border tab

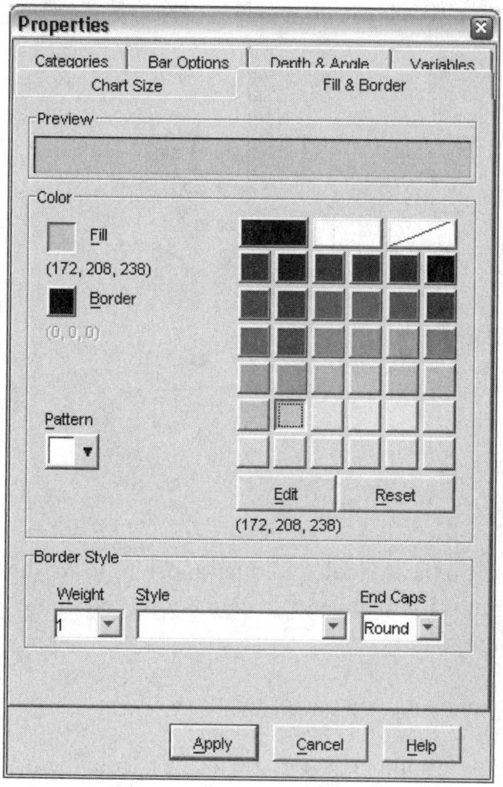

▶ Click Apply.

The bars in the chart are now light blue.

Figure 8-11
Edited bar chart showing blue bars

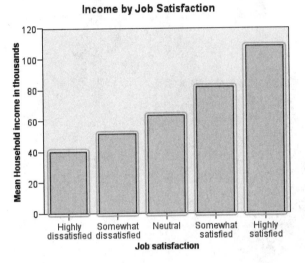

Formatting Numbers in Tick Labels

Notice that the numbers on the *y* axis are scaled in thousands. To make the chart more attractive and easier to interpret, we will change the number format in the tick labels and then edit the axis title appropriately.

▶ Select the *y* axis tick labels by clicking any one of them.

▶ To reopen the Properties window (if you closed it previously), from the menus choose:
  Edit
    Properties

Note: From here on, we assume that the Properties window is open. If you have closed the Properties window, follow the previous step to reopen it. You can also use the keyboard shortcut Ctrl+T to reopen the window.

▶ Click the Number Format tab.

▶ You do not want the tick labels to display decimal places, so type 0 in the Decimal Places text box.

▶ Type 0.001 in the Scaling Factor text box. The scaling factor is the number by which the Chart Editor divides the displayed number. Because 0.001 is a fraction, dividing by it will *increase* the numbers in the tick labels by 1,000. Thus, the numbers will no longer be in thousands; they will be unscaled.

▶ Select Display Digit Grouping. Digit grouping uses a character (specified by your computer's locale) to mark each thousandth place in the number.

Figure 8-12
Number Format tab

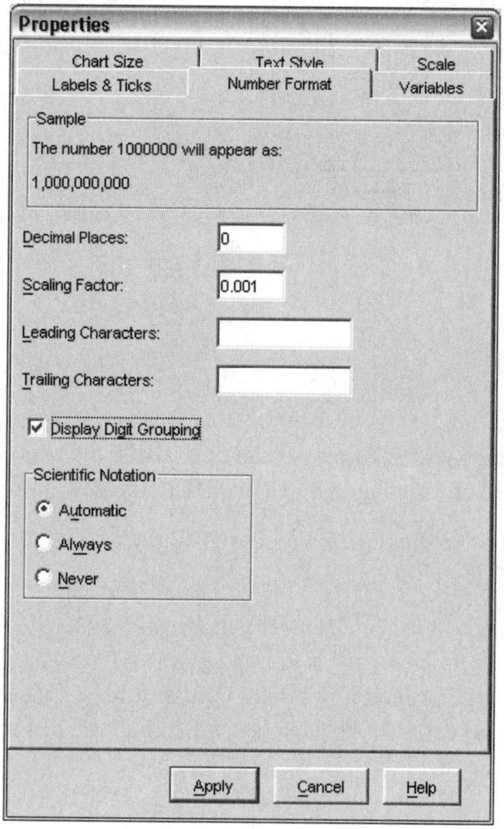

▶ Click Apply.

The tick labels reflect the new number formatting: There are no decimal places, the numbers are no longer scaled, and each thousandth place is specified with a character.

Figure 8-13
Edited bar chart showing new number format

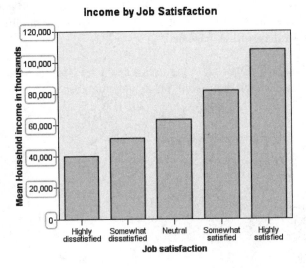

Editing Text

Now that you have changed the number format of the tick labels, the axis title is no longer accurate. Next, you will change the axis title to reflect the new number format.

Note: You do not need to open the Properties window to edit text. You can edit text directly on the chart.

▶ Click the *y* axis title to select it.

▶ Click the axis title again to start edit mode. While in edit mode, the Chart Editor positions any rotated text horizontally. It also displays a flashing red bar cursor (not shown in the example).

▶ Delete the following text:

in thousands

▶ Press Enter to exit edit mode and update the axis title. The axis title now accurately describes the contents of the tick labels.

Figure 8-14
Bar chart showing edited y axis title

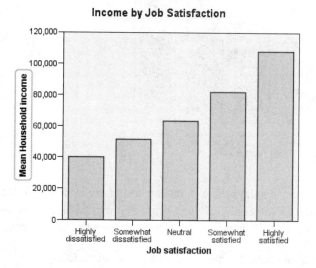

Displaying and Editing Data Value Labels

Another common task is to show the exact values associated with the data elements (which are bars in this example). These values are displayed in data labels.

▶ From the Chart Editor menus choose:
Elements
  Show Data Labels

154

Chapter 8

Figure 8-15
Bar chart showing data value labels

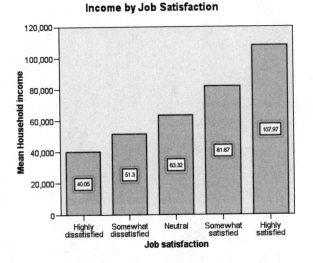

Each bar in the chart now displays the exact mean household income. Notice that the units are in thousands, so you could use the Number Format tab again to change the scaling factor. Instead, we will change the statistic shown in the data values.

▶ With the data value labels selected, click the Data Value Labels tab.

The Data Value Labels tab allows you to change various label properties. In this case, you are going to change the content of the labels so that they display the percentage of respondents in each category.

▶ Move *Mean income* from the Displayed list to the Not Displayed list.

▶ Move *Percent* from the Not Displayed list to the Displayed list.

Figure 8-16
Data Value Labels tab

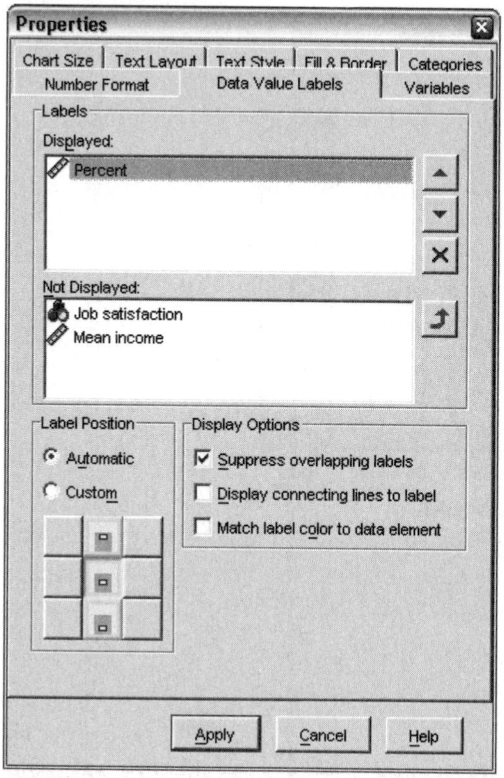

▶ Click Apply to update the data value labels.

The bars now display the percentage of cases in each *Job satisfaction* category.

Figure 8-17
Bar chart showing percentages

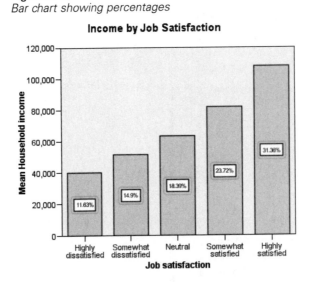

Income by Job Satisfaction

Using Templates

If you make a number of routine changes to your charts, you can use a chart template to reduce the time needed to create and edit charts. A chart template saves the attributes of a specific chart. You can then apply the template when creating or editing a chart.

We will save the current chart as a template and then apply that template while creating a new chart.

▶ From the menus choose:
File
  Save Chart Template...

The Save Chart Template dialog box allows you to specify which chart attributes you want to include in the template.

If you expand any of the items in the tree view, you can see which specific attributes can be saved with the chart. For example, if you expand the Scale axes portion of the tree, you can see all of the attributes of data value labels that the template will include. You can select any attribute to include it in the template.

▶ Select All settings to include all of the available chart attributes in the template.

You can also enter a description of the template. This description will be visible when you apply the template.

Figure 8-18
Save Chart Template dialog box

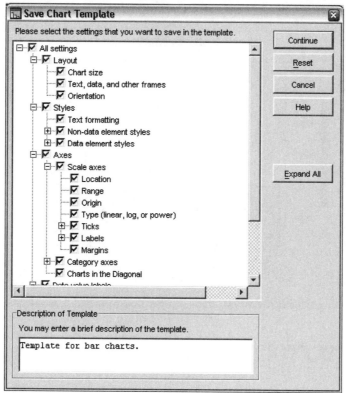

▶ Click Continue.

▶ In the Save Template dialog box, specify a location and filename for the template.

▶ When you are finished, click Save.

You can apply the template when you create a chart or in the Chart Editor. In the following example, we will apply it while creating a chart.

▶ Close the Chart Editor. The updated bar chart is shown in the Viewer.

Figure 8-19
Updated bar chart in Viewer

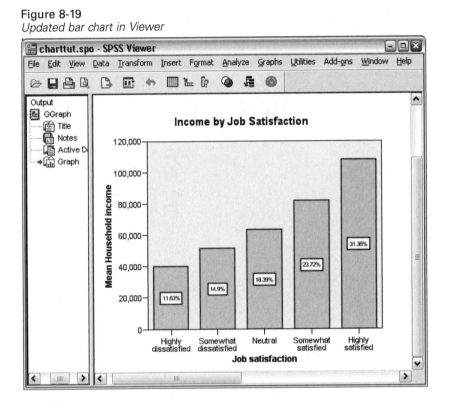

▶ From the Viewer menus choose:
Graphs
  Chart Builder...

The Chart Builder dialog box "remembers" the variables that you entered when you created the original chart. However, here you will create a slightly different chart to see how applying a template formats a chart.

▶ Remove *Job satisfaction* from the *x* axis by dragging it from the drop zone back to the Variables list. You can also click the drop zone and press Delete.

▶ Right-click *Level of education* in the Variables list and choose Ordinal.

▶ Drag *Level of education* from the Variables list to the *x* axis drop zone.

Because the title is now inaccurate, we are going to delete it.

▶ Click Element Properties.

▶ In the Edit Properties of list, select Title 1.

▶ Click the red *X* located to the right of the Edit Properties of list.

▶ Click Apply.

Now we are going to specify the template to apply to the new chart.

▶ Click Options.

▶ In the Templates group in the Options dialog box, click Add.

▶ In the Find Template Files dialog box, locate the template file that you previously saved using the Save Chart Template dialog box.

▶ Select that file and click Open.

Figure 8-20
Options dialog box with template

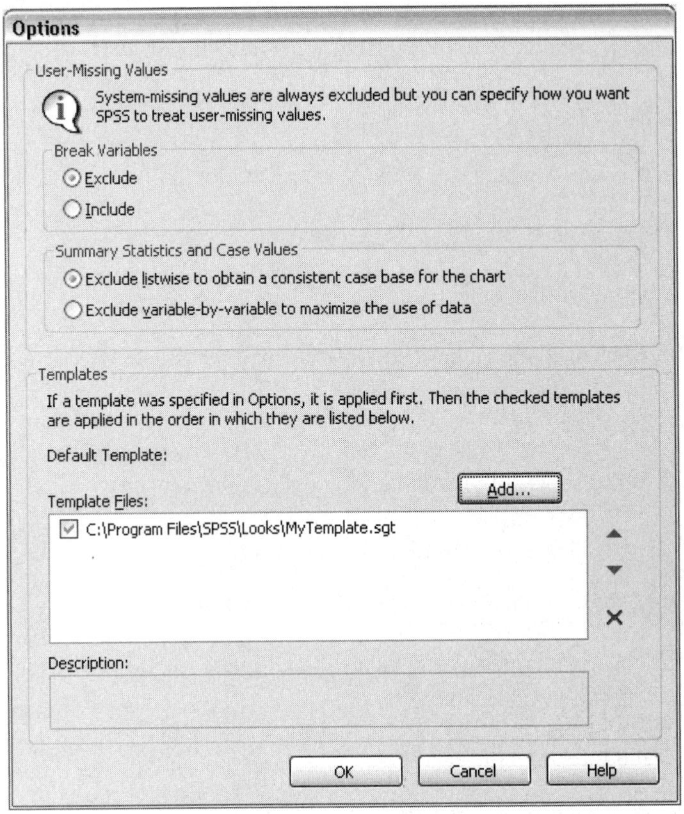

The Options dialog box displays the file path of the template you selected.

(Our example shows the path *C:\Program Files\SPSS\Looks\My Template.sgt.*)

▶ Click OK to close the Options dialog box.

Figure 8-21
Chart Builder with completed drop zones

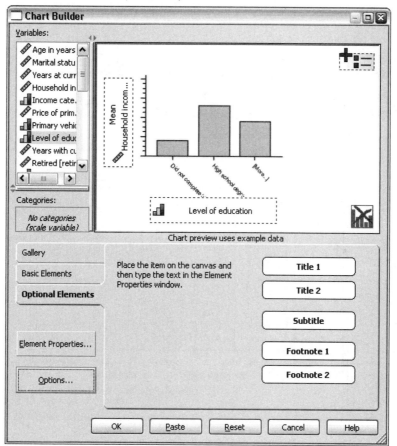

▶ Click OK on the Chart Builder dialog box to create the chart and apply the template.

The formatting in the new chart matches the formatting in the chart that you previously created and edited. Although the variables on the *x* axis are different, the charts otherwise resemble each other. Notice that the title from the previous chart was preserved in the template, even though you deleted the title in the Chart Builder.

If you want to apply templates after you've created a chart, you can do it in the Chart Editor (from the File menu, choose Apply Chart Template).

Figure 8-22
Updated bar chart in Viewer

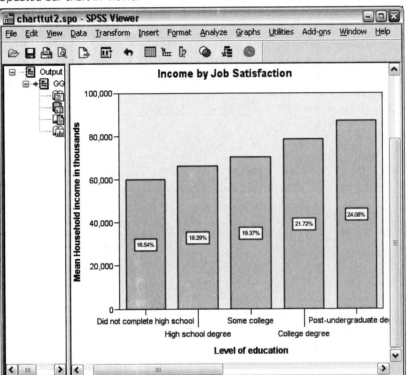

Defining Chart Options

In addition to using templates to format charts, you can use the SPSS options to control various aspects of how charts are created.

▶ From the Data Editor or Viewer menus choose:
Edit
  Options...

The Options dialog box contains many settings for configuring SPSS. Click the Charts tab to see the available options.

Figure 8-23
Charts tab in Options dialog box

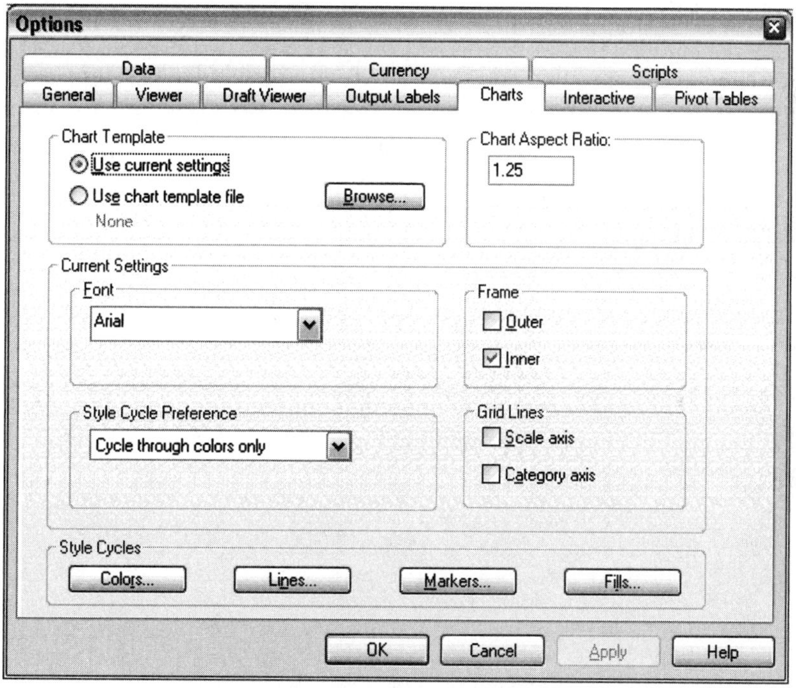

The options control how a chart is created. For each new chart, you can specify:

- Whether to use the current settings or a template.
- The width-to-height ratio (aspect ratio).
- If you're not using a template, the settings to use for formatting.
- The style cycles for data elements.

Style cycles allow you to specify the style of data elements in new charts. In this example, we'll look at the details for the color style cycle.

▶ Click Colors to open the Data Element Colors dialog box.

For a simple chart, the Chart Editor uses one style that you specify. For grouped charts, the Chart Editor uses a set of styles that it cycles through for each group (category) in the chart.

▶ Select Simple Charts.

▶ Select the light green color, which is third from the right in the third row from the bottom.

Figure 8-24
Data Element Colors dialog box

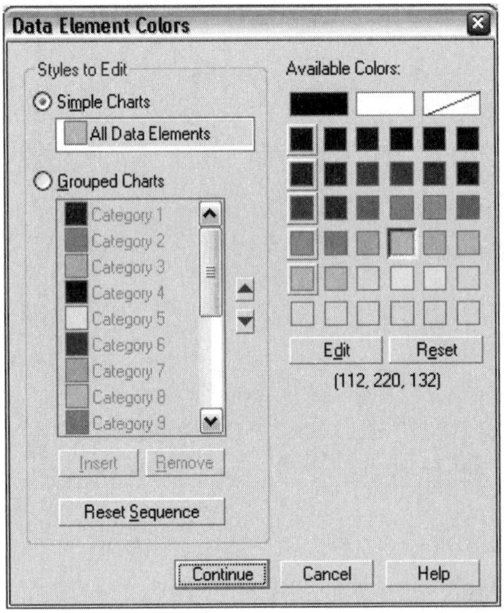

▶ Click Continue.

▶ In the Options dialog box, click OK to save the color style cycle changes.

The data elements in any new simple charts will now be light green.

▶ From the Data Editor or Viewer menus choose:
Graphs
  Chart Builder...

The Chart Builder displays the last chart you created, but this time with light green bars. Remember that this chart had a template associated with it. We no longer want to use that template.

▶ Click Options.

▶ Deselect (uncheck) the template that you added previously. Note that you could also click the red *X* to delete the template. By deselecting rather than deleting, you keep the template available to use at another time.

▶ Click OK to create the chart.

The bars in the new chart are light green.

Figure 8-25
Updated bar chart in Viewer

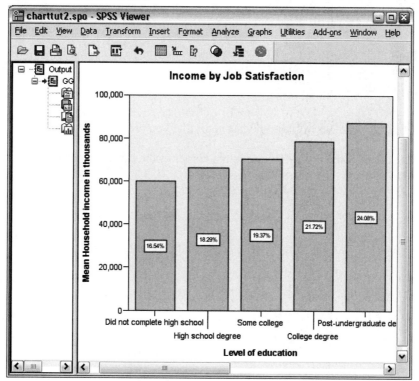

Other Examples

Now we will create and edit a pie chart and a grouped scatterplot to explore other editing capabilities, including:

- Hiding categories.
- Moving text.
- Converting a chart to another chart type.
- Adding a fit line to a scatterplot.
- Identifying points in a scatterplot.

Pie Chart

First, we will create a simple pie chart that shows how many respondents have Internet service at home. This example uses the data file *demo.sav*.

► From the menus choose:
Graphs
  Chart Builder...

The Chart Builder displays the last chart that you created.

► Click the Clear icon in the lower right corner of the canvas to clear the previous chart.

► Click the Gallery tab and click Pie/Polar.

► Drag the pie chart onto the canvas.

► Right-click *Internet* in the Variables list and choose Nominal.

▶ Drag *Internet* from the Variables list to the Slice By drop zone. This is the variable that defines the number of pie slices.

Figure 8-26
Completed Chart Builder for pie chart

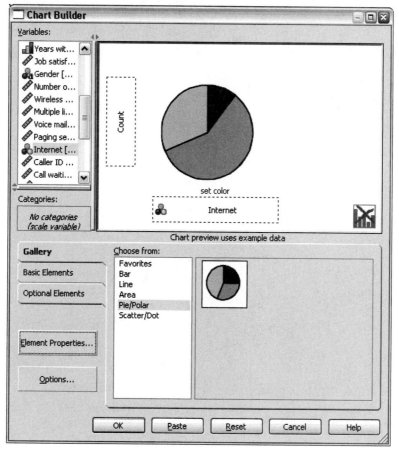

When you create charts, they do not show the *missing* category by default. You want to display this category to make sure that the number of cases with missing values is not excessive.

▶ Click Options.

Figure 8-27
Options dialog box

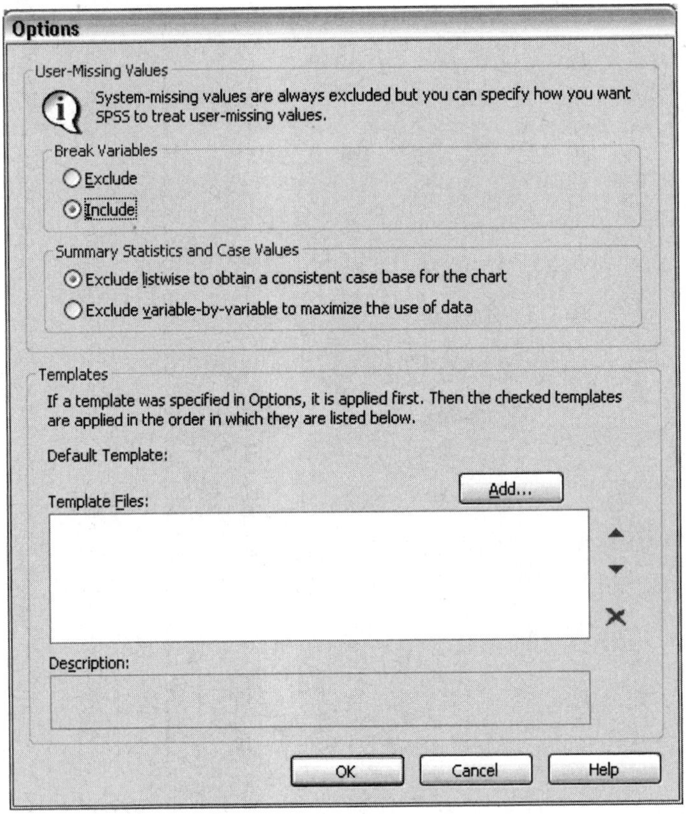

Options

User-Missing Values

System-missing values are always excluded but you can specify how you want SPSS to treat user-missing values.

Break Variables

○ Exclude
⊙ Include

Summary Statistics and Case Values

⊙ Exclude listwise to obtain a consistent case base for the chart
○ Exclude variable-by-variable to maximize the use of data

Templates

If a template was specified in Options, it is applied first. Then the checked templates are applied in the order in which they are listed below.

Default Template:

Template Files: Add...

Description:

OK   Cancel   Help

▶ In the Break Variables group, select Include, and then click OK.

▶ Click OK in the Chart Builder dialog box to create the pie chart.

Figure 8-28
Pie chart in Viewer

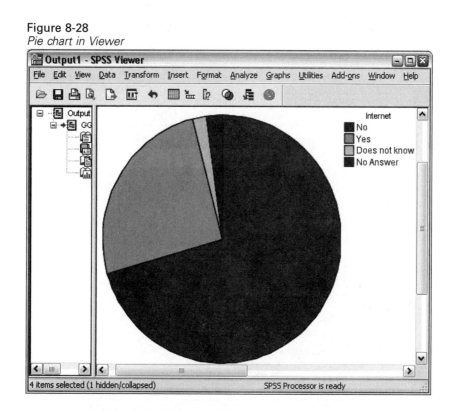

The pie chart reveals that most respondents do not have Internet service at home. From the chart, it appears that only about a quarter of the respondents have home Internet service.

For this pie chart, you will:

- Add a title.

- Remove the categories of missing data.

- Display percentages for the two remaining categories.

- Move the data labels and connect them to the slices with lines.

- Convert the pie chart to a bar chart.

▶ Double-click the pie chart to open it in the Chart Editor.

Figure 8-29
Pie chart in the Chart Editor

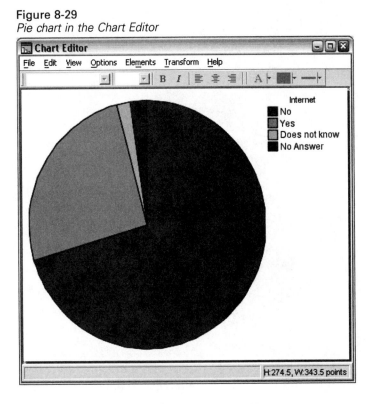

Adding a Chart Title

First, you will add a title to the chart. Because the title applies to the entire chart and not to a specific chart element, you don't need to select anything to add the title.

▶ From the menus choose:
Options
  Title

The Chart Editor adds the word *Title* above the chart and enlarges the chart to accommodate the title. Like footnotes, the title is in a text frame that you can move to another position in the chart.

Figure 8-30
Pie chart showing default title

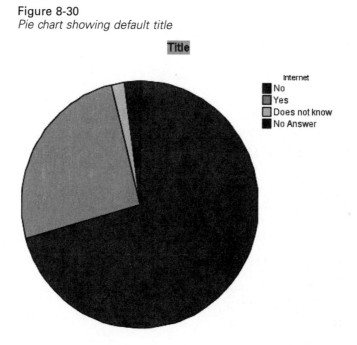

▶ Type Home Internet Service over the highlighted default title.

Note: If you clicked elsewhere in the Chart Editor, the default title may no longer be highlighted. The text frame around the title may be selected instead. If this is the case, click the title again to start edit mode. You can then double-click the title text and begin typing.

Figure 8-31
Pie chart showing edited title

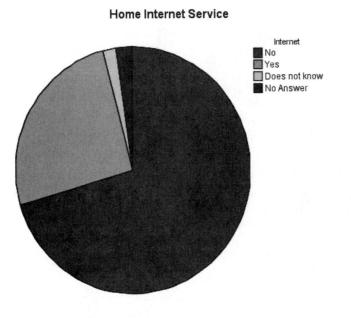

Home Internet Service

Modifying Chart Categories

Next, you will remove the small category of missing data.

▶ In the Chart Editor, select the pie chart.

▶ Click the Categories tab in the Properties window. (If the window is closed, choose Properties from the Edit menu or press Ctrl+T.)

▶ Move *Does not know* and *No Answer* from the Order list to the Excluded list.

Figure 8-32
Categories tab

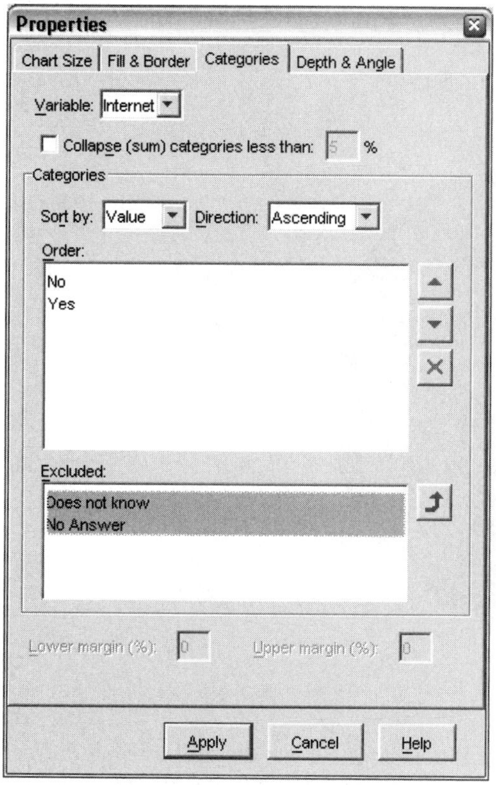

▶ Click Apply.

The pie chart now displays only the *No* and *Yes* categories for home Internet service.

Figure 8-33
Pie chart with excluded missing categories

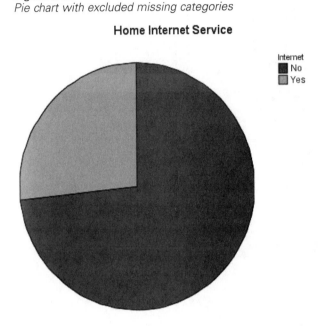

Home Internet Service

The pie chart clearly shows that most respondents do not have Internet service at home; it appears that about three-quarters of the respondents are in the *No* category. However, it might be useful to see the exact percentages.

Changing Data Value Label Content and Location

▶ In the Chart Editor, select the pie chart.

▶ From the menus choose:
Elements
   Show Data Labels

The pie chart now displays labels of counts. We need to change the counts to percentages.

▶ Click the Data Value Labels tab in the Properties window.

▶ Move *Count* from the Displayed list to the Not Displayed list.

▶ Move *Percent* from the Not Displayed list to the Displayed list.

Figure 8-34
Data Value Labels tab

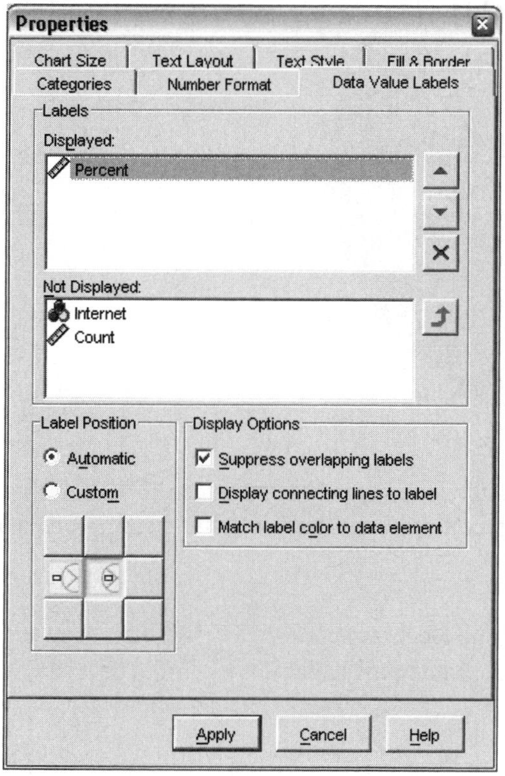

▶ Click Apply.

Percentages are now displayed in the pie slices.

Figure 8-35
Pie chart showing percentages

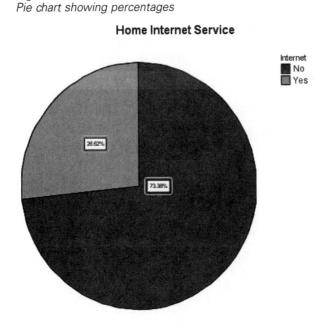

Home Internet Service

The percentages are based on the two categories displayed (73.4 + 26.6 = 100). If you put the categories containing missing values back into the pie, the percentages will change.

You may not want the data value labels to appear in the pie slices. You can move them and add connecting lines to their respective slices.

▶ Return to the Data Value Labels tab.

▶ In the Label Position group, click Custom.

▶ Click the icon showing the labels outside the slices.

▶ Select Display connecting lines to label.

▶ Click Apply.

Data value labels now appear outside the slices with lines connecting the labels to their respective slices.

Figure 8-36
Labels outside of pie chart

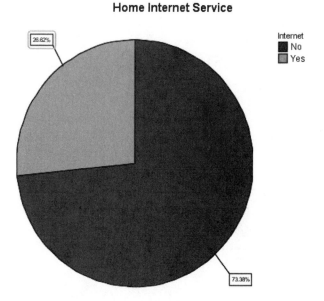

Home Internet Service

Converting a Chart

Finally, you will convert the pie chart into a bar chart.

▶ From the menus choose:
Transform
  Simple Bar

The Chart Editor converts the chart and displays the result. The bar chart retains the title and data labels. However, the bars are the same color rather than the two different colors shown in the pie chart. Because the tick labels identify the categories, two colors are unnecessary.

Figure 8-37
Bar chart converted from pie chart

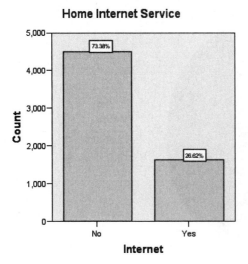

Grouped Scatterplot

In this example, we will create a scatterplot using the data file *car_sales.sav*. The scatterplot will show fuel efficiency against curb weight.

▶ From the Data Editor menus choose:
Graphs
  Chart Builder...

▶ Click the Gallery tab and click Scatter/Dot.

▶ Drag the Grouped Scatter icon onto the canvas.

For a scatterplot, you usually define a scale variable for each axis. Put the dependent variable on the *y* axis and the independent variable on the *x* axis.

▶ Drag *Curb weight* from the Variables list to the *x* axis drop zone.

▶ Drag *Fuel efficiency* from the Variables list to the *y* axis drop zone.

▶ Drag *Vehicle type* from the Variables list to the grouping drop zone, which is located in the upper right corner of the canvas.

The grouping drop zone is used when another variable splits data elements into groups. In this case, the grouping variable creates distinct groups of points in the resulting scatterplot. In a bar chart, the grouping variable is used for features such as clustering and stacking.

Figure 8-38
Chart Builder dialog box

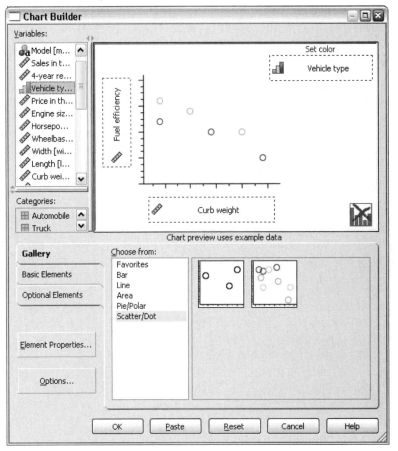

▶ Click OK to create the scatterplot.

▶ In the Viewer, double-click the resulting scatterplot to display it in the Chart Editor.

Figure 8-39
Scatterplot in the Chart Editor

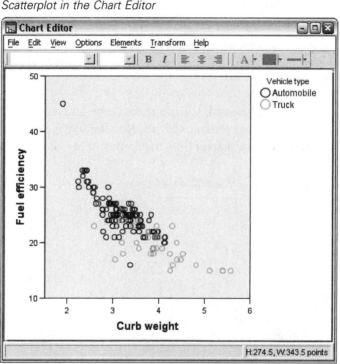

Selecting Elements in Grouped Charts

Selecting an element in grouped charts (that is, charts with categories) differs from selecting an element in simple charts. You can tell that a chart is grouped because it includes, by default, a legend for the groups.

You can hide the legend if necessary (choose Hide Legend from the Options menu). However, the legend provides important information for interpreting the chart. You can also use the legend to easily select groups.

There are general rules for selecting elements in grouped charts:

■ When no data elements are selected, click any data element to select all data elements.

- When all data elements are selected, click a data element in a specific group to select all data elements in the group. You can then click a data element in another group to select its associated group.

- You can also click the legend entry for a group to select only that group.

- When a group of data elements is selected, click a data element in the group to select only that data element. You can then select another data element in the chart by clicking it.

With specific data elements selected, you can make changes to only the selected data elements. So, you could change the color of data elements in a group. In this example, we changed the fill color of the marker from transparent to the same color as its border.

Figure 8-40
Scatterplot showing one case with a different fill color

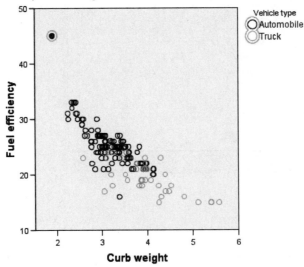

Adding a Fit Line

With scatterplots, you often want to add a fit line to the chart. For elements that, like fit lines, apply to the whole chart, it doesn't matter what is selected in the chart. You can make the change or add the item to the whole chart. If you were creating a fit line for only one or more specific groups, you would need to make specific selections.

▶ From the menus choose:
Elements
  Fit Line at Total

Figure 8-41
Scatterplot showing fit line at total

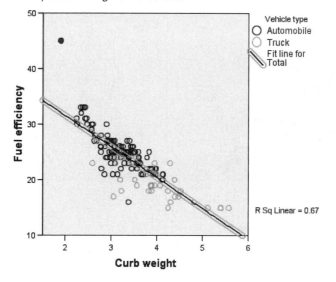

The Chart Editor adds the fit line to the chart. It also displays the coefficient of determination, R^2. R^2 is the proportion of the variability in the dependent variable (y axis) that can be explained by the independent variable (x axis). In this example, we can say that 67% of the variability in *Fuel efficiency* values can be explained by *Curb weight*. The closer R^2 is to 1, the better the line fits the data.

Displaying Data Value Labels for Specific Points

Previously, in the bar and pie chart examples, we displayed the data value labels for all data elements. With scatterplots, displaying data value labels for all markers is not usually useful. There are too many data elements, making viewing of the associated data value labels difficult.

▶ For example, from the menus choose:
Elements
  Show Data Labels

The results are cluttered and of limited use.

Figure 8-42
Scatterplot showing all data value labels

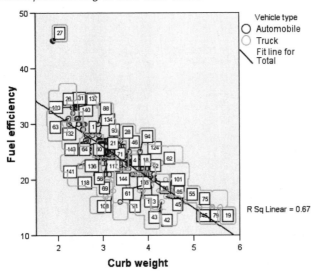

Not every marker in the scatterplot has a displayed data value label. This occurs because of a Data Value Labels tab setting that suppresses overlapping labels. This setting helps to eliminate some of the clutter, but you don't have control over which particular data value labels are displayed. You could turn off this setting to display all data value labels, even the overlapping ones. However, many labels would become unreadable. A better option would be to display data value labels for only those points that you want to highlight.

First, hide the current data value labels.

▶ From the menus choose:
Elements
  Hide Data Labels

Next, turn on data label mode, which allows you to click any data element and automatically display its data value label.

▶ From the menus choose:
Elements
  Data Label Mode

The cursor changes to indicate that you are in data label mode.

▶ Click the marker for which we previously changed the fill color.

Figure 8-43
Scatterplot with one marker showing a data value label

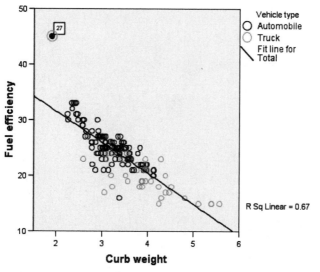

The Chart Editor now displays the data value label for only that marker. You can repeat this process for each marker that needs a data value label. You can also hide a data value label by clicking it while in data label mode.

The data value label contains the case number. You can change the contents of the data value label on the Data Value Labels tab.

▶ When you are finished labeling the data elements, exit data label mode. From the menus choose:
Elements
  Data Label Mode

Working with Syntax

SPSS provides a powerful command language that allows you to save and automate many common tasks. It also provides some functionality not found in the menus and dialog boxes. Most commands are accessible from the menus and dialog boxes. However, some commands and options are available only by using the command language. The command language also allows you to save your jobs in a syntax file so that you can repeat your analysis at a later date or run it in an automated job with the Production Facility.

A command syntax file is simply a text file that contains SPSS commands. You can open a syntax window and type commands directly, but it is often easier to let the dialog boxes do some or all of the work for you.

The examples in this chapter use the data file *demo.sav*.

Note: Command syntax is not available with the Student Version.

Pasting Syntax

The easiest way to create syntax is to use the Paste button located on most dialog boxes.

▶ Open *demo.sav* for use in this example.

▶ From the menus choose:
Analyze
  Descriptive Statistics
    Frequencies...

The Frequencies dialog box opens.

Figure 9-1
Frequencies dialog box

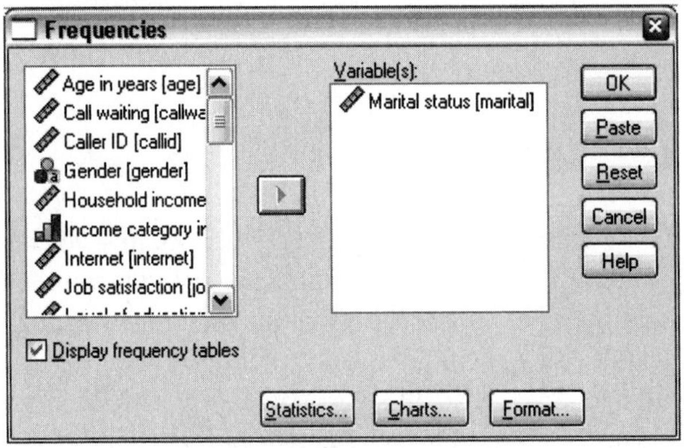

▶ Select *Marital status [marital]* in the source list.

▶ Click the arrow button to move the variable to the Variable(s) list.

▶ Click Charts.

▶ In the Charts dialog box, select Bar charts.

▶ In the Chart Values group, select Percentages.

▶ Click Continue.

▶ Click Paste to copy the syntax created as a result of the dialog box selections to the Syntax Editor.

Figure 9-2
Frequencies syntax

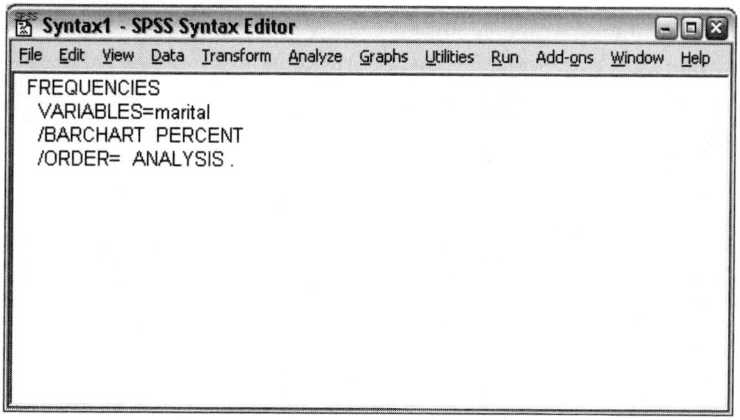

```
FREQUENCIES
 VARIABLES=marital
 /BARCHART PERCENT
 /ORDER= ANALYSIS .
```

You can use this syntax alone, add it to a larger syntax file, or refer to it in a Production Facility job.

▶ To run the syntax currently displayed, from the menus choose:
Run
  Current

Editing Syntax

In the syntax window, you can edit the syntax. For example, you could change the subcommand /BARCHART to display frequencies instead of percentages. (A subcommand is indicated by a slash.)

Figure 9-3
Modified syntax

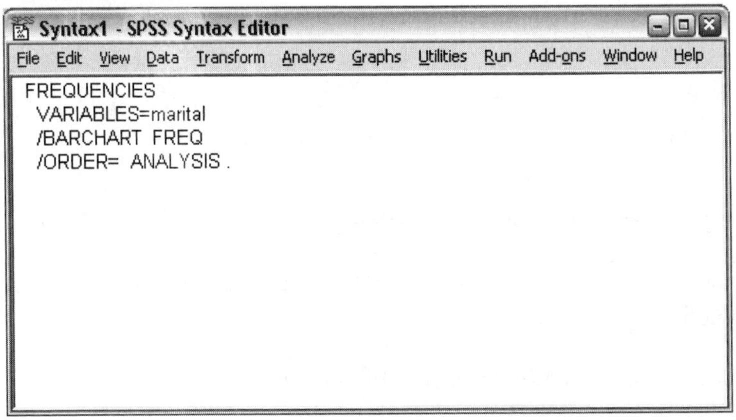

To find out what subcommands and keywords are available for the current command, press the F1 key. This takes you directly to the command syntax reference information for the current command.

Figure 9-4
FREQUENCIES syntax help

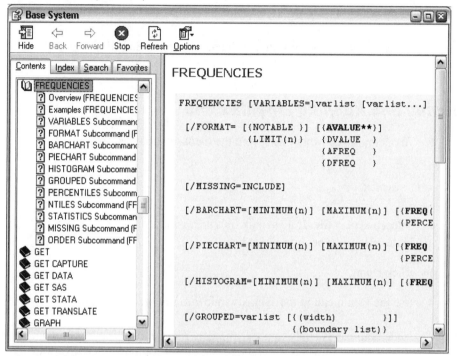

Typing Command Syntax

You can type syntax into a syntax window that is already open, or you can open a new syntax window by choosing:

File
  New
    Syntax...

Saving Syntax

To save a syntax file, from the menus choose:

File
  Save

or

File
  Save As...

Either action opens the standard dialog box for saving files.

Opening and Running a Syntax File

▶ To open a saved syntax file, from the menus choose:

File
  Open
    Syntax...

▶ Select a syntax file. If no syntax files are displayed, make sure Syntax (*.sps) is selected in the Files of type drop-down list.

▶ Click Open.

▶ Use the Run menu in the syntax window to run the commands.

If the commands apply to a specific data file, the data file must be opened before running the commands, or you must include a command that opens the data file. You can paste this type of command from the dialog boxes that open data files.

Modifying Data Values

The data you start with may not always be organized in the most useful manner for your analysis or reporting needs. For example, you may want to:

- Create a categorical variable from a scale variable.

- Combine several response categories into a single category.

- Create a new variable that is the computed difference between two existing variables.

- Calculate the length of time between two dates.

This chapter uses the data file *demo.sav*.

Creating a Categorical Variable from a Scale Variable

Several categorical variables in the data file *demo.sav* are, in fact, derived from scale variables in that data file. For example, the variable *inccat* is simply *income* grouped into four categories. This categorical variable uses the integer values 1–4 to represent the following income categories: less than 25, 25–49, 50–74, and 75 or higher.

To create the categorical variable *inccat*:

▶ From the menus in the Data Editor window choose:
Transform
  Visual Bander...

Figure 10-1
Initial Visual Bander dialog box

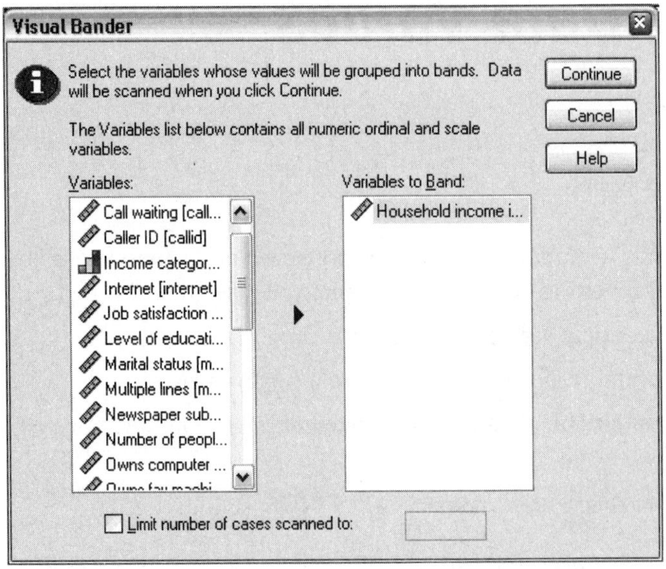

In the initial Visual Bander dialog box, you select the scale and/or ordinal variables for which you want to create new, banded variables. **Banding** means taking two or more contiguous values and grouping them into the same category.

Since the Visual Bander relies on actual values in the data file to help you make good banding choices, it needs to read the data file first. Since this can take some time if your data file contains a large number of cases, this initial dialog box also allows you to limit the number of cases to read ("scan"). This is not necessary for our sample data file. Even though it contains more than 6,000 cases, it does not take long to scan that number of cases.

▶ Drag and drop *Household income in thousands [income]* from the Variables list into the Variables to Band list, and then click Continue.

Figure 10-2
Main Visual Bander dialog box

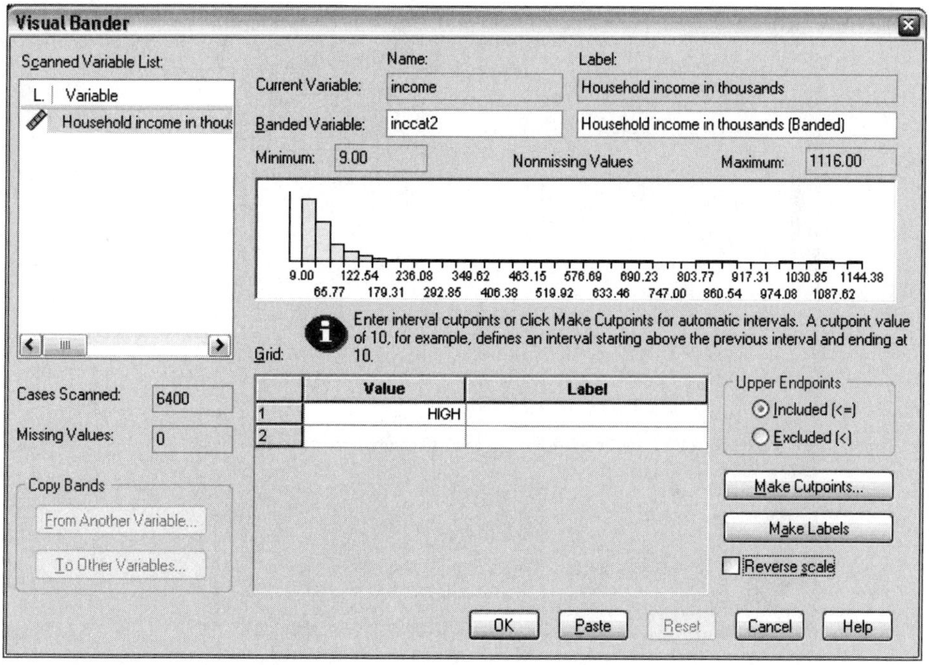

> In the main Visual Bander dialog box, select *Household income in thousands [income]* in the Scanned Variable List.

A histogram displays the distribution of the selected variable (which in this case is highly skewed).

> Enter inccat2 for the new banded variable name and Income category [in thousands] for the variable label.

> Click Make Cutpoints.

Figure 10-3
Visual Bander Cutpoints dialog box

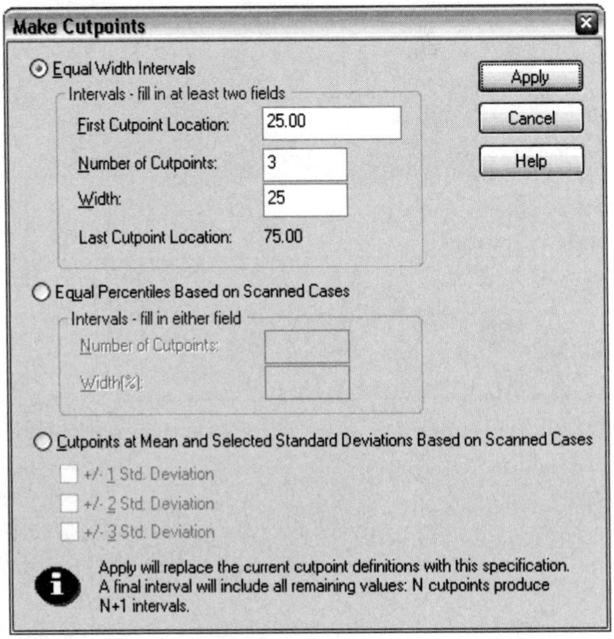

▶ Select Equal Width Intervals.

▶ Enter 25 for the first cutpoint location, 3 for the number of cutpoints, and 25 for the width.

The number of banded categories is one greater than the number of cutpoints. So in this example, the new banded variable will have four categories, with the first three categories each containing ranges of 25 (thousand) and the last one containing all values above the highest cutpoint value of 75 (thousand).

▶ Click Apply.

Figure 10-4
Main Visual Bander dialog box with defined cutpoints

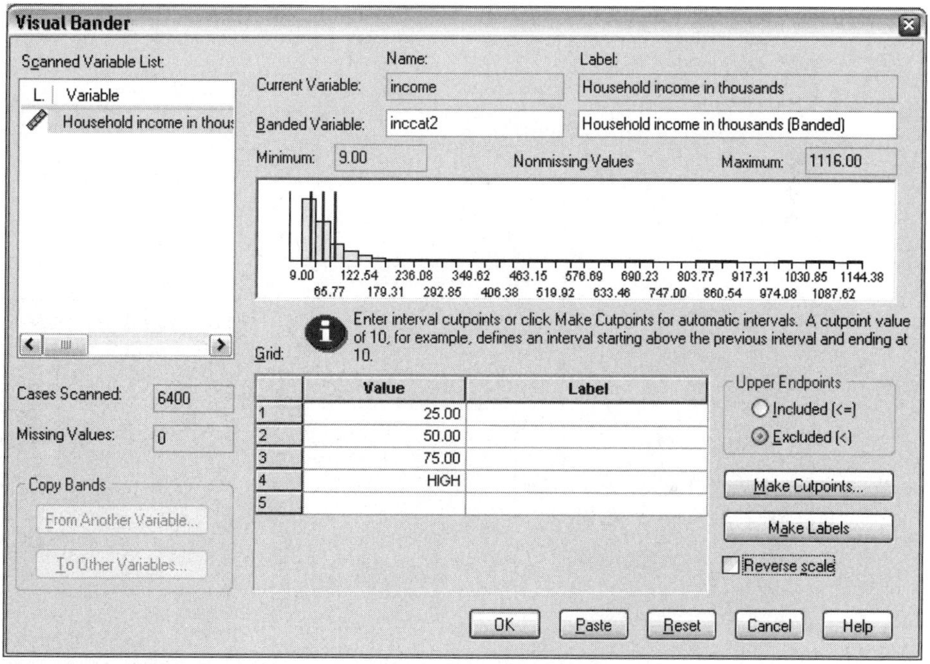

The values now displayed in the grid represent the defined cutpoints, which are the upper endpoints of each category. Vertical lines in the histogram also indicate the locations of the cutpoints.

By default, these cutpoint values are included in the corresponding categories. For example, the first value of 25 would include all values less than or equal to 25. But in this example, we want categories that correspond to less than 25, 25–49, 50–74, and 75 or higher.

▶ In the Upper Endpoints group, select Excluded (<).

▶ Then click Make Labels.

Figure 10-5
Automatically generated value labels

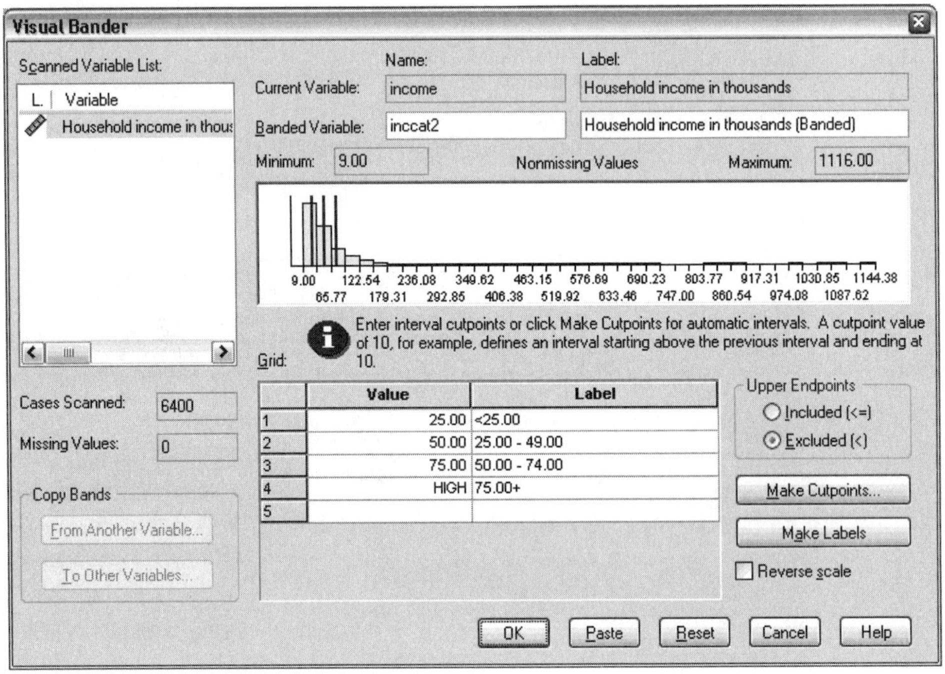

This automatically generates descriptive value labels for each category. Since the actual values assigned to the new banded variable are simply sequential integers starting with 1, the value labels can be very useful.

You can also manually enter or change cutpoints and labels in the grid, change cutpoint locations by dragging and dropping the cutpoint lines in the histogram, and delete cutpoints by dragging cutpoint lines off of the histogram.

▶ Click OK to create the new, banded variable.

The new variable is displayed in the Data Editor. Since the variable is added to the end of the file, it is displayed in the far right column in Data View and in the last row in Variable View.

Figure 10-6
New variable displayed in Data Editor

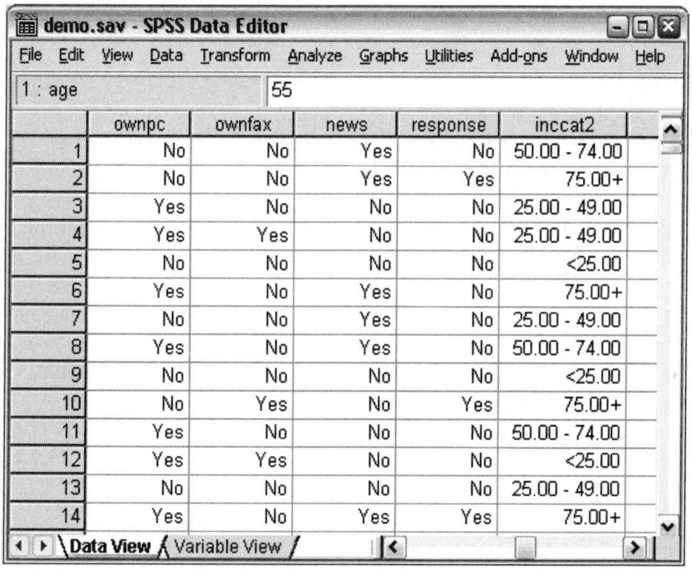

demo.sav - SPSS Data Editor

File Edit View Data Transform Analyze Graphs Utilities Add-ons Window Help

1 : age 55

	ownpc	ownfax	news	response	inccat2
1	No	No	Yes	No	50.00 - 74.00
2	No	No	Yes	Yes	75.00+
3	Yes	No	No	No	25.00 - 49.00
4	Yes	Yes	No	No	25.00 - 49.00
5	No	No	No	No	<25.00
6	Yes	No	Yes	No	75.00+
7	No	No	Yes	No	25.00 - 49.00
8	Yes	No	Yes	No	50.00 - 74.00
9	No	No	No	No	<25.00
10	No	Yes	No	Yes	75.00+
11	Yes	No	No	No	50.00 - 74.00
12	Yes	Yes	No	No	<25.00
13	No	No	No	No	25.00 - 49.00
14	Yes	No	Yes	Yes	75.00+

Data View / Variable View

Computing New Variables

Using a wide variety of mathematical functions, you can compute new variables based on highly complex equations. In this example, however, we will simply compute a new variable that is the difference between the values of two existing variables.

The data file *demo.sav* contains a variable for the respondent's current age and a variable for the number of years at current job. It does not, however, contain a variable for the respondent's age at the time he or she started that job. We can create a new variable that is the computed difference between current age and number of years at current job, which should be the approximate age at which the respondent started that job.

▶ From the menus in the Data Editor window choose:

Transform
   Compute...

▶ For Target Variable, enter jobstart.

▶ Select *Age in years [age]* in the source variable list and click the arrow button to copy it to the Numeric Expression text box.

▶ Click the minus (–) button on the calculator pad in the dialog box (or press the minus key on the keyboard).

▶ Select *Years with current employer [employ]* and click the arrow button to copy it to the expression.

Figure 10-7
Compute Variable dialog box

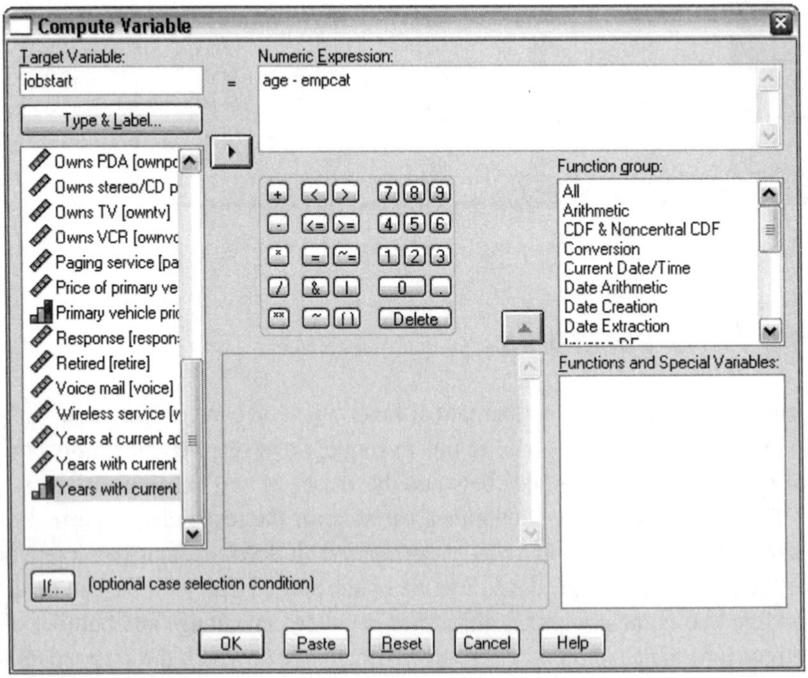

Note: Be careful to select the correct employment variable. There is also a recoded categorical version of the variable, which is *not* what you want. The numeric expression should be *age-employ*, not *age-empcat*.

▶ Click OK to compute the new variable.

The new variable is displayed in the Data Editor. Since the variable is added to the end of the file, it is displayed in the far right column in Data View and in the last row in Variable View.

Figure 10-8
New variable displayed in Data Editor

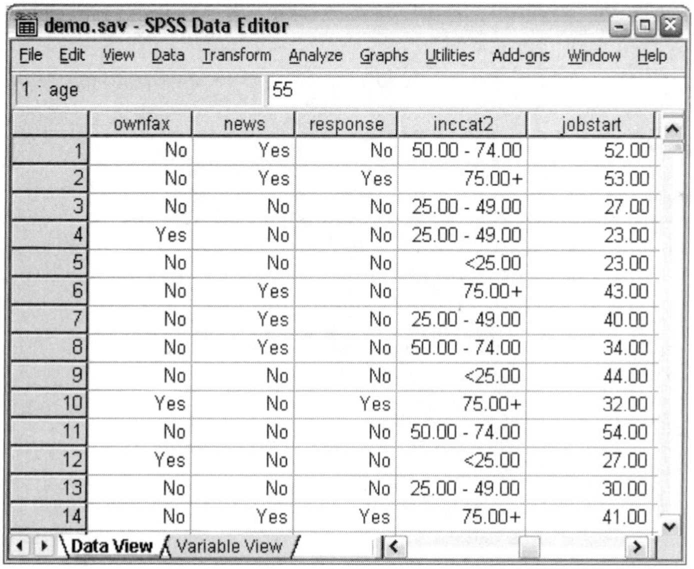

	ownfax	news	response	inccat2	jobstart
1	No	Yes	No	50.00 - 74.00	52.00
2	No	Yes	Yes	75.00+	53.00
3	No	No	No	25.00 - 49.00	27.00
4	Yes	No	No	25.00 - 49.00	23.00
5	No	No	No	<25.00	23.00
6	No	Yes	No	75.00+	43.00
7	No	Yes	No	25.00 - 49.00	40.00
8	No	Yes	No	50.00 - 74.00	34.00
9	No	No	No	<25.00	44.00
10	Yes	No	Yes	75.00+	32.00
11	No	No	No	50.00 - 74.00	54.00
12	Yes	No	No	<25.00	27.00
13	No	No	No	25.00 - 49.00	30.00
14	No	Yes	Yes	75.00+	41.00

Using Functions in Expressions

You can also use predefined functions in expressions. More than 70 built-in functions are available, including:

- Arithmetic functions
- Statistical functions
- Distribution functions
- Logical functions

- Date and time aggregation and extraction functions

- Missing-value functions

- Cross-case functions

- String functions

Figure 10-9
Compute Variable dialog box displaying function grouping

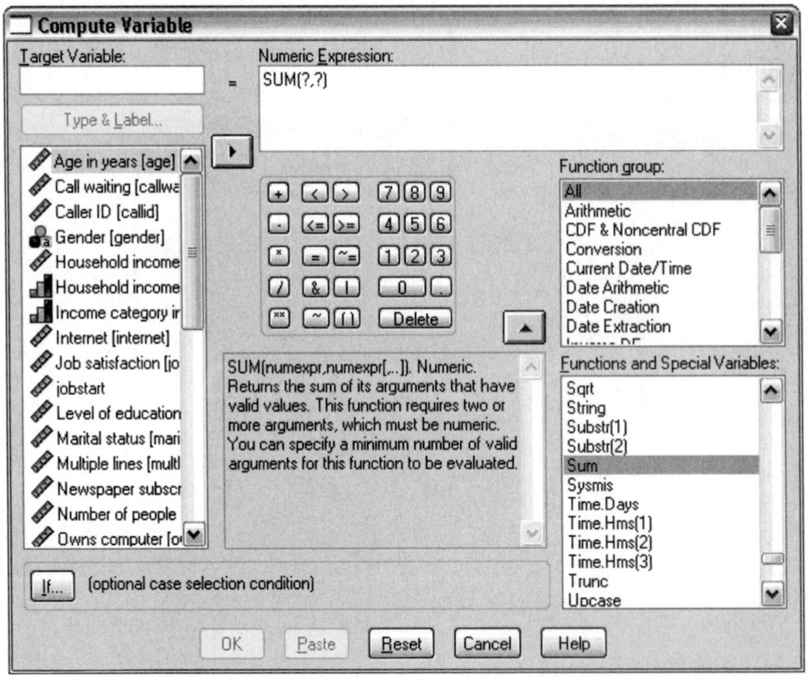

Functions are organized into logically distinct groups, such as a group for arithmetic operations and another for computing statistical metrics. For convenience, a number of commonly used system variables, such as $TIME (current date and time), are also included in appropriate function groups. A brief description of the currently selected function (in this case, SUM) or system variable is displayed in a reserved area in the Compute Variable dialog box.

Pasting a function into an expression. To paste a function into an expression:

▶ Position the cursor in the expression at the point where you want the function to appear.

▶ Select the appropriate group from the Function group list. The group labeled All provides a listing of all available functions and system variables.

▶ Double-click the function in the Functions and Special Variables list (or select the function and click the arrow adjacent to the Function group list).

The function is inserted into the expression. If you highlight part of the expression and then insert the function, the highlighted portion of the expression is used as the first argument in the function.

Editing a function in an expression. The function is not complete until you enter the arguments, represented by question marks in the pasted function. The number of question marks indicates the minimum number of arguments required to complete the function.

▶ Highlight the question mark(s) in the pasted function.

▶ Enter the arguments. If the arguments are variable names, you can paste them from the variable list.

Using Conditional Expressions

You can use conditional expressions (also called logical expressions) to apply transformations to selected subsets of cases. A conditional expression returns a value of true, false, or missing for each case. If the result of a conditional expression is true, the transformation is applied to that case. If the result is false or missing, the transformation is not applied to the case.

To specify a conditional expression:

▶ Click If in the Compute Variable dialog box. This opens the If Cases dialog box.

Figure 10-10
If Cases dialog box

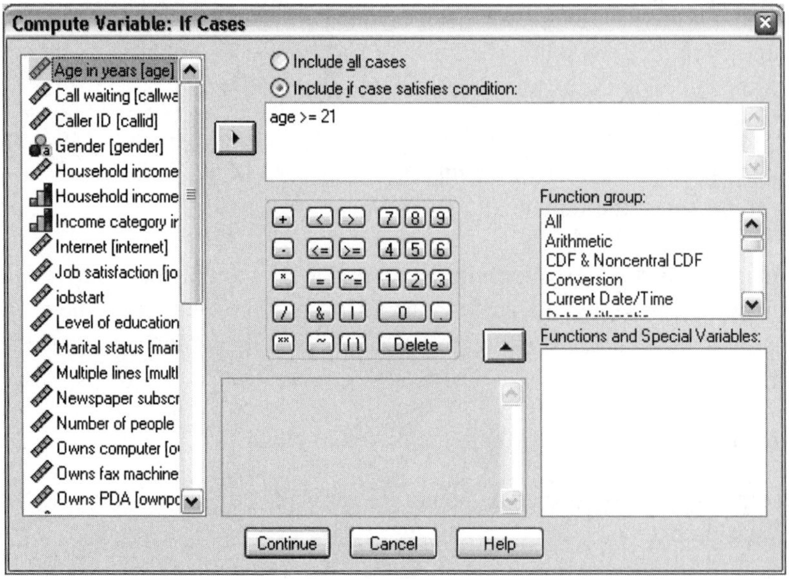

▶ Select Include if case satisfies condition.

▶ Enter the conditional expression.

Most conditional expressions contain at least one relational operator, as in:

```
age>=21
```

or

```
income*3<100
```

In the first example, only cases with a value of 21 or greater for *Age [age]* are selected. In the second example, *Household income in thousands [income]* multiplied by 3 must be less than 100 for a case to be selected.

You can also link two or more conditional expressions using logical operators, as in:

```
age>=21 | ed>=4
```

or

```
income*3<100 & ed=5
```

In the first example, cases that meet either the *Age [age]* condition or the *Level of education [ed]* condition are selected. In the second example, both the *Household income in thousands [income]* and *Level of education [ed]* conditions must be met for a case to be selected.

Working with Dates and Times

A number of tasks commonly performed with dates and times can be easily accomplished using the Date and Time Wizard. Using this wizard, you can:

■ Create a date/time variable from a string variable containing a date or time.

■ Construct a date/time variable by merging variables containing different parts of the date or time.

■ Add or subtract values from date/time variables, including adding or subtracting two date/time variables.

■ Extract a part of a date or time variable; for example, the day of month from a date/time variable which has the form mm/dd/yyyy.

The examples in this section use the data file *upgrade.sav*.

To use the Date and Time Wizard:

▶ From the menus choose:
Transform
  Date/Time...

Figure 10-11
Date and Time Wizard Introduction Screen

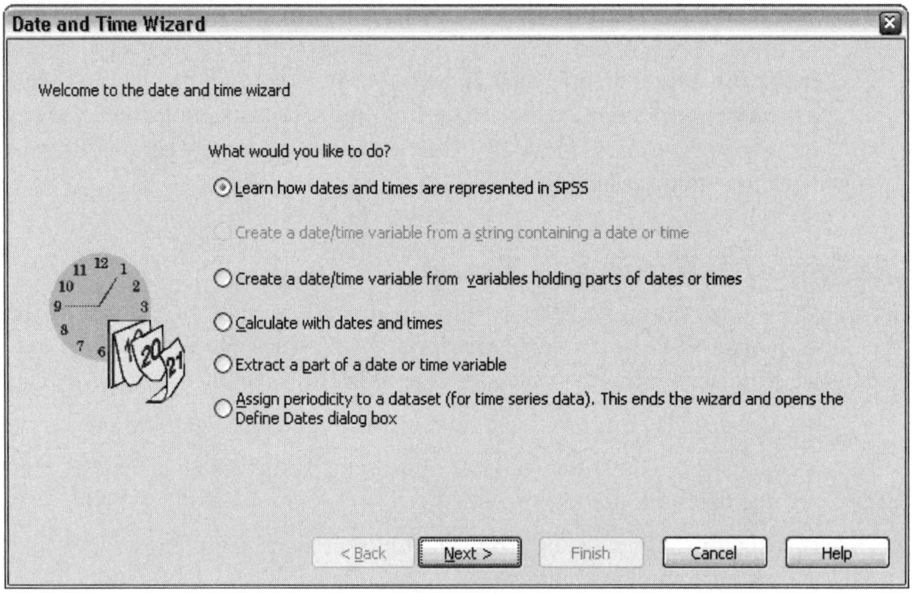

The introduction screen of the Date and Time Wizard presents you with a set of general tasks. Tasks that do not apply to the current data are disabled. For example, the data file *upgrade.sav* doesn't contain any string variables, so the task to create a date variable from a string is disabled.

If you're new to dates and times in SPSS, you can select Learn how dates and times are represented in SPSS and click Next. This leads to a screen that provides a brief overview of date/time variables, and a link, through the Help button, to more detailed information.

Calculating the Length of Time Between Two Dates

One of the most common tasks involving dates is calculating the length of time between two dates. As an example, consider a software company interested in analyzing purchases of upgrade licenses by determining the number of years since each customer last purchased an upgrade. The data file *upgrade.sav* contains a variable for the date on which each customer last purchased an upgrade, but not the number of years since that purchase. A new variable that is the length of time, in years, between the date of the last upgrade and the date of the next product release will provide a measure of this quantity.

To calculate the length of time between two dates:

▶ Select Calculate with dates and times on the introduction screen of the Date and Time Wizard and click Next.

Figure 10-12
Calculating the length of time between two dates: Step 1

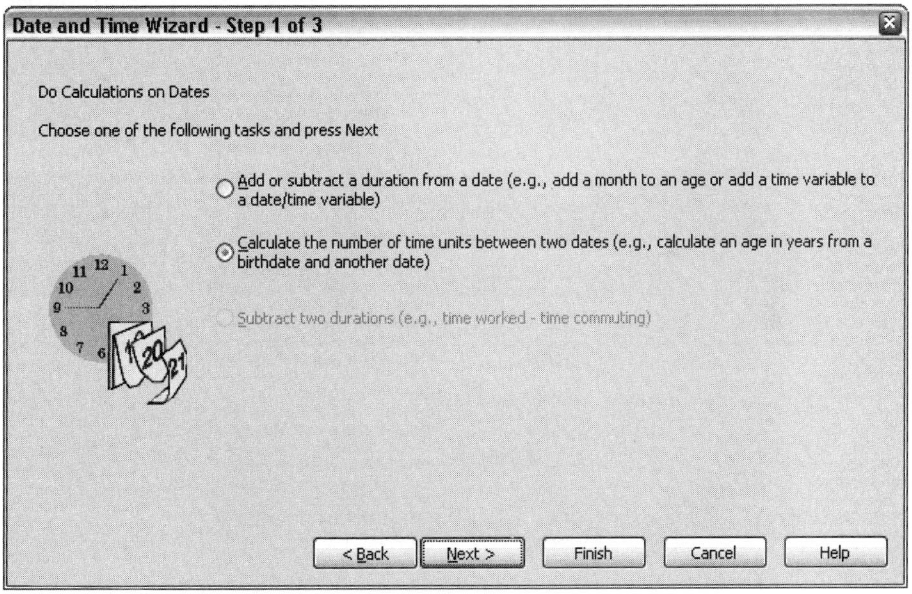

▶ Select Calculate the number of time units between two dates and click Next.

Figure 10-13
Calculating the length of time between two dates: Step 2

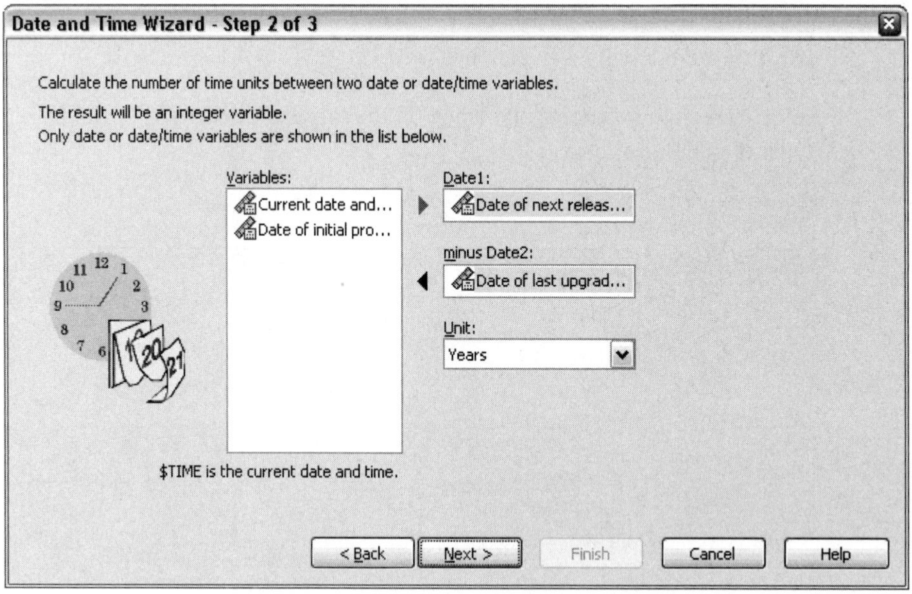

▶ Select *Date of next release* for Date1.

▶ Select *Date of last upgrade* for Date2.

▶ Leave the Unit drop-down list at the default of Years.

▶ Click Next.

Figure 10-14
Calculating the length of time between two dates: Step 3

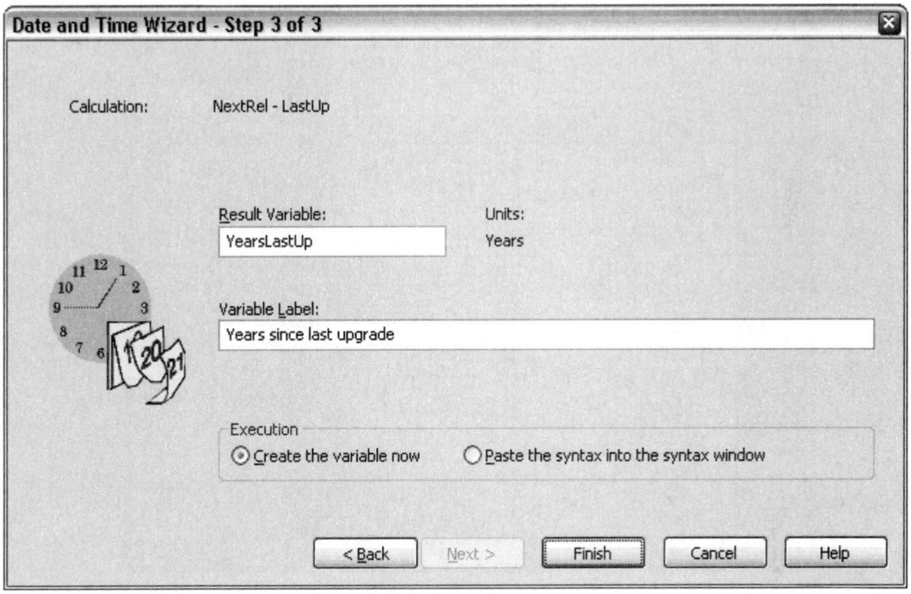

▶ Enter *YearsLastUp* for the name of the result variable. Result variables cannot have the same name as an existing variable.

▶ Enter *Years since last upgrade* as the label for the result variable. Variable labels for result variables are optional.

▶ Leave the default selection of Create the variable now, and click Finish to create the new variable.

The new variable *YearsLastUp*, displayed in the Data Editor, is the integer number of years between the two dates. Fractional parts of a year have been truncated.

Figure 10-15

New variable displayed in Data Editor

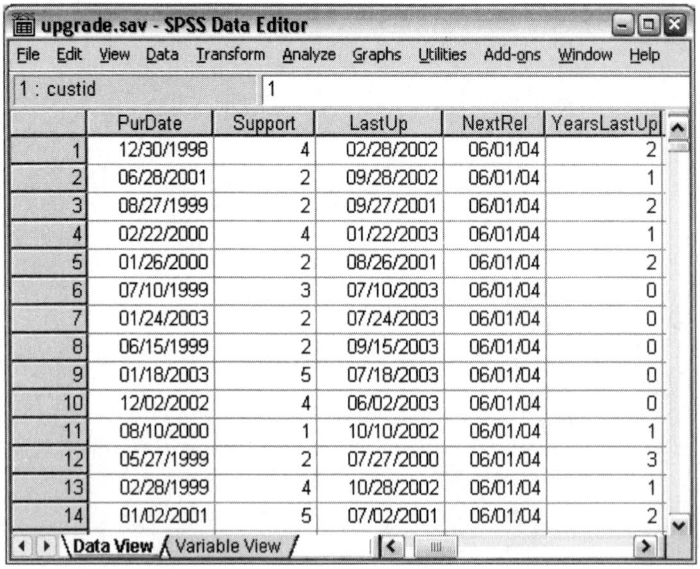

	PurDate	Support	LastUp	NextRel	YearsLastUp
1	12/30/1998	4	02/28/2002	06/01/04	2
2	06/28/2001	2	09/28/2002	06/01/04	1
3	08/27/1999	2	09/27/2001	06/01/04	2
4	02/22/2000	4	01/22/2003	06/01/04	1
5	01/26/2000	2	08/26/2001	06/01/04	2
6	07/10/1999	3	07/10/2003	06/01/04	0
7	01/24/2003	2	07/24/2003	06/01/04	0
8	06/15/1999	2	09/15/2003	06/01/04	0
9	01/18/2003	5	07/18/2003	06/01/04	0
10	12/02/2002	4	06/02/2003	06/01/04	0
11	08/10/2000	1	10/10/2002	06/01/04	1
12	05/27/1999	2	07/27/2000	06/01/04	3
13	02/28/1999	4	10/28/2002	06/01/04	1
14	01/02/2001	5	07/02/2001	06/01/04	2

Adding a Duration to a Date

You can add or subtract durations, such as 10 days or 12 months, to a date. Continuing with the example of the software company from the previous section, consider determining the date on which each customer's initial tech support contract ends. The data file *upgrade.sav* contains a variable for the number of years of contracted support and a variable for the initial purchase date. You can then determine the end date of the initial support by adding years of support to the purchase date.

To add a duration to a date:

▶ Select Calculate with dates and times on the introduction screen of the Date and Time Wizard and click Next.

▶ Select Add or subtract a duration from a date and click Next.

Figure 10-16
Adding a duration to a date: Step 2

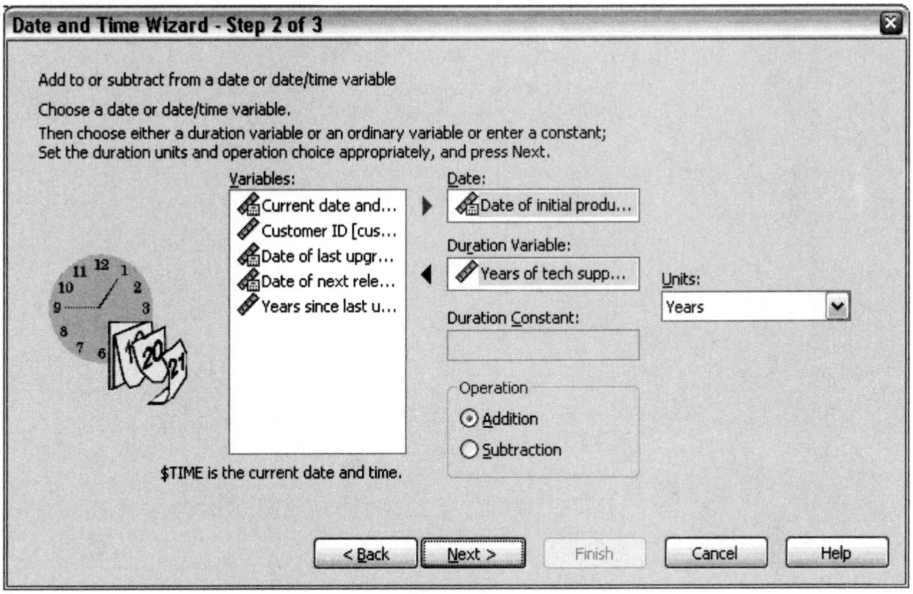

▶ Select *Date of initial product license* for Date.

▶ Select *Years of tech support* for the Duration Variable.

Since *Years of tech support* is simply a numeric variable, you need to indicate the units to use when adding this variable as a duration.

▶ Select Years from the Units drop-down list.

▶ Click Next.

Figure 10-17
Adding a duration to a date: Step 3

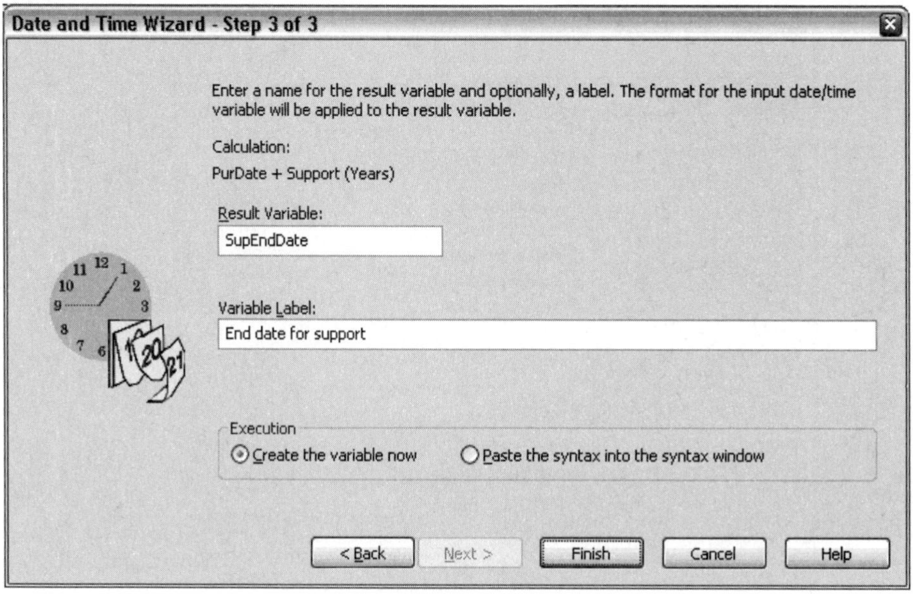

▶ Enter *SupEndDate* for the name of the result variable. Result variables cannot have the same name as an existing variable.

▶ Enter *End date for support* as the label for the result variable. Variable labels for result variables are optional.

▶ Click Finish to create the new variable.

The new variable is displayed in the Data Editor.

Figure 10-18
New variable displayed in Data Editor

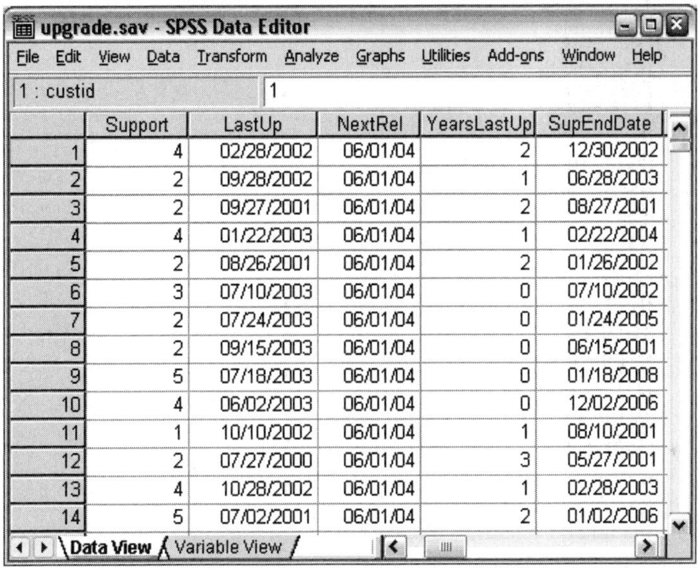

	Support	LastUp	NextRel	YearsLastUp	SupEndDate
1	4	02/28/2002	06/01/04	2	12/30/2002
2	2	09/28/2002	06/01/04	1	06/28/2003
3	2	09/27/2001	06/01/04	2	08/27/2001
4	4	01/22/2003	06/01/04	1	02/22/2004
5	2	08/26/2001	06/01/04	2	01/26/2002
6	3	07/10/2003	06/01/04	0	07/10/2002
7	2	07/24/2003	06/01/04	0	01/24/2005
8	2	09/15/2003	06/01/04	0	06/15/2001
9	5	07/18/2003	06/01/04	0	01/18/2008
10	4	06/02/2003	06/01/04	0	12/02/2006
11	1	10/10/2002	06/01/04	1	08/10/2001
12	2	07/27/2000	06/01/04	3	05/27/2001
13	4	10/28/2002	06/01/04	1	02/28/2003
14	5	07/02/2001	06/01/04	2	01/02/2006

upgrade.sav - SPSS Data Editor

File Edit View Data Transform Analyze Graphs Utilities Add-ons Window Help

1 : custid   1

Data View   Variable View

Sorting and Selecting Data

Data files are not always organized in the ideal form for your specific needs. To prepare data for analysis, you can select from a wide range of file transformations, including the ability to:

- **Sort data.** You can sort cases based on the value of one or more variables.
- **Select subsets of cases.** You can restrict your analysis to a subset of cases or perform simultaneous analyses on different subsets.

The examples in this chapter use the data file *demo.sav*.

Sorting Data

Sorting cases (sorting rows of the data file) is often useful and sometimes necessary for certain types of analysis.

To reorder the sequence of cases in the data file based on the value of one or more sorting variables:

▶ From the menus choose:
Data
    Sort Cases...

The Sort Cases dialog box is displayed.

Figure 11-1
Sort Cases dialog box

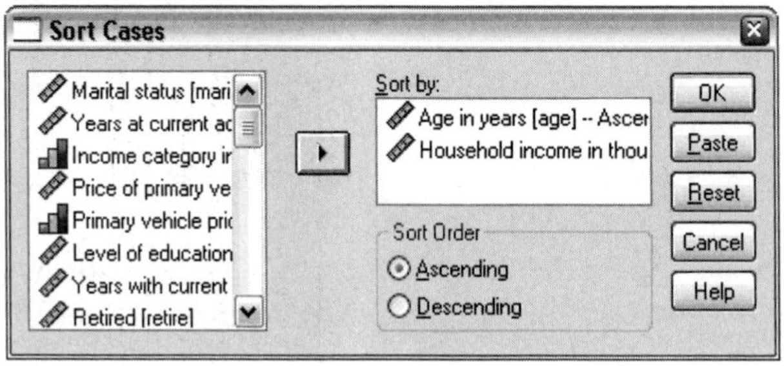

▶ Add the *Age in years [age]* and *Household income in thousands [income]* variables to the Sort by list.

If you select multiple sort variables, the order in which they appear on the Sort by list determines the order in which cases are sorted. In this example, based on the entries in the Sort by list, cases will be sorted by the value of *Household income in thousands [income]* within categories of *Age in years [age]*. For string variables, uppercase letters precede their lowercase counterparts in sort order (for example, the string value *Yes* comes before *yes* in the sort order).

Split-File Processing

To split your data file into separate groups for analysis:

▶ From the menus choose:
Data
  Split File...

The Split File dialog box is displayed.

Figure 11-2
Split File dialog box

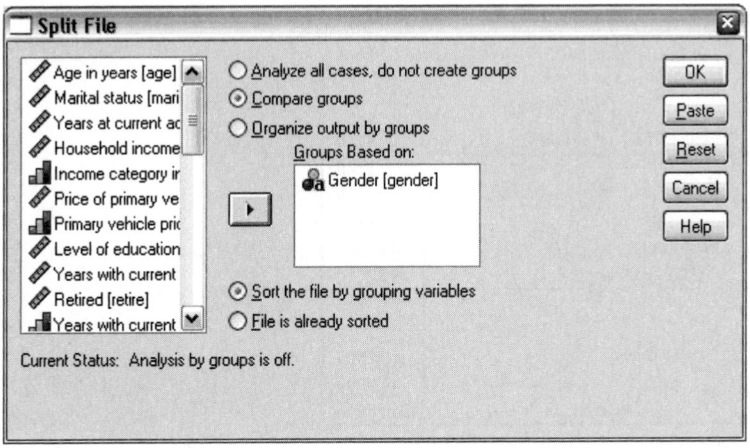

▶ Select Compare groups or Organize output by groups. (The examples following these steps show the differences between these two options.)

▶ Select *Gender [gender]* to split the file into separate groups for these variables.

You can use numeric, short string, and long string variables as grouping variables. A separate analysis is performed for each subgroup that is defined by the grouping variables. If you select multiple grouping variables, the order in which they appear on the Groups Based on list determines the manner in which cases are grouped.

If you select Compare groups, results from all split-file groups will be included in the same table(s), as in the following table of summary statistics that is generated by the Frequencies procedure.

Figure 11-3
Split File output with single pivot table

Statistics

Household income in thousands

Female	N	Valid	3179
		Missing	0
	Mean		68.7798
	Median		44.0000
	Std. Deviation		75.73510
Male	N	Valid	3221
		Missing	0
	Mean		70.1608
	Median		45.0000
	Std. Deviation		81.56216

If you select Organize output by groups and run the Frequencies procedure, two pivot tables are created: one table for females and one table for males.

Figure 11-4
Split File output with pivot table for females

Statistics^a

Household income in thousands

N	Valid	3179
	Missing	0
Mean		68.7798
Median		44.0000
Std. Deviation		75.73510

a. Gender = Female

Figure 11-5
Split File output with pivot table for males

Statistics[a]

Household income in thousands

N	Valid	3221
	Missing	0
Mean		70.1608
Median		45.0000
Std. Deviation		81.56216

a. Gender = Male

Sorting Cases for Split-File Processing

The Split File procedure creates a new subgroup each time it encounters a different value for one of the grouping variables. Therefore, it is important to sort cases based on the values of the grouping variables before invoking split-file processing.

By default, Split File automatically sorts the data file based on the values of the grouping variables. If the file is already sorted in the proper order, you can save processing time if you select File is already sorted.

Turning Split-File Processing On and Off

After you invoke split-file processing, it remains in effect for the rest of the session unless you turn it off.

- **Analyze all cases.** This option turns split-file processing off.

- **Compare groups** and **Organize output by groups.** This option turns split-file processing on.

If split-file processing is in effect, the message Split File on appears on the status bar at the bottom of the application window.

Selecting Subsets of Cases

You can restrict your analysis to a specific subgroup based on criteria that include variables and complex expressions. You can also select a random sample of cases. The criteria used to define a subgroup can include:

- Variable values and ranges
- Date and time ranges
- Case (row) numbers
- Arithmetic expressions
- Logical expressions
- Functions

To select a subset of cases for analysis:

▶ From the menus choose:
Data
  Select Cases...

This opens the Select Cases dialog box.

Figure 11-6
Select Cases dialog box

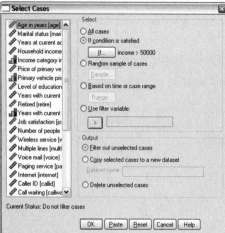

Selecting Cases Based on Conditional Expressions

To select cases based on a conditional expression:

► Select If condition is satisfied and click If in the Select Cases dialog box.

This opens the Select Cases If dialog box.

Figure 11-7
Select Cases If dialog box

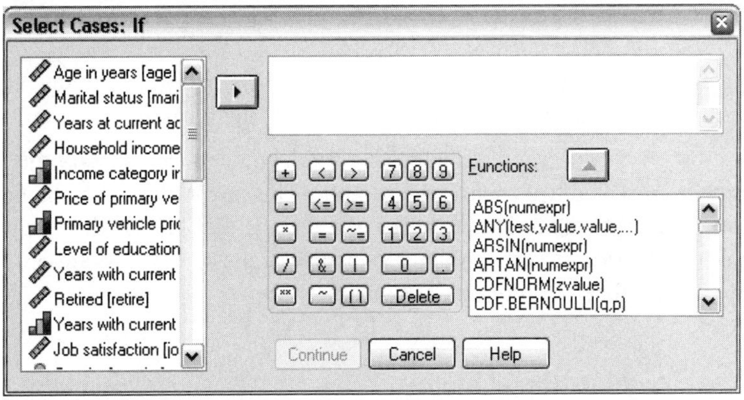

The conditional expression can use existing variable names, constants, arithmetic operators, logical operators, relational operators, and functions. You can type and edit the expression in the text box just like text in an output window. You can also use the calculator pad, variable list, and function list to paste elements into the expression. For more information, see "Using Conditional Expressions" in Chapter 10 on p. 203.

Selecting a Random Sample

To obtain a random sample:

► Select Random sample of cases in the Select Cases dialog box.

► Click Sample.

This opens the Select Cases Random Sample dialog box.

Figure 11-8
Select Cases Random Sample dialog box

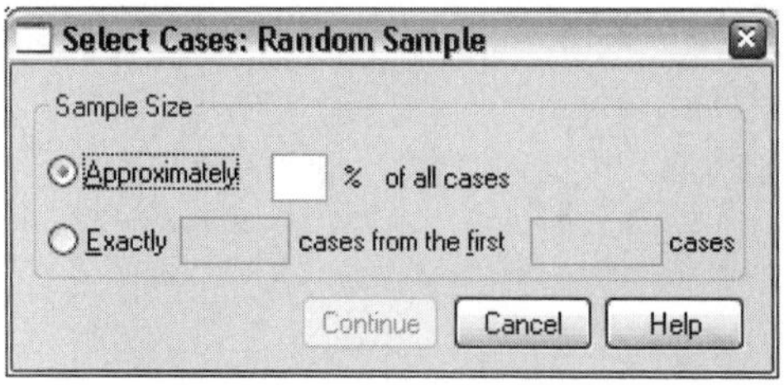

You can select one of the following alternatives for sample size:

■ **Approximately.** A user-specified percentage. This option generates a random sample of approximately the specified percentage of cases.

■ **Exactly.** A user-specified number of cases. You must also specify the number of cases from which to generate the sample. This second number should be less than or equal to the total number of cases in the data file. If the number exceeds the total number of cases in the data file, the sample will contain proportionally fewer cases than the requested number.

Selecting a Time Range or Case Range

To select a range of cases based on dates, times, or observation (row) numbers:

▶ Select Based on time or case range and click Range in the Select Cases dialog box.

This opens the Select Cases Range dialog box, in which you can select a range of observation (row) numbers.

Figure 11-9
Select Cases Range dialog box

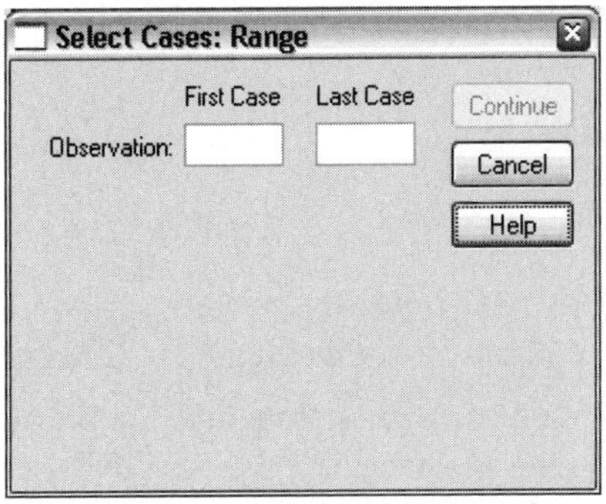

- **First Case.** Enter the starting date and/or time values for the range. If no date variables are defined, enter the starting observation number (row number in the Data Editor, unless Split File is on). If you do not specify a Last Case value, all cases from the starting date/time to the end of the time series are selected.

- **Last Case.** Enter the ending date and/or time values for the range. If no date variables are defined, enter the ending observation number (row number in the Data Editor, unless Split File is on). If you do not specify a First Case value, all cases from the beginning of the time series up to the ending date/time are selected.

For time series data with defined date variables, you can select a range of dates and/or times based on the defined date variables. Each case represents observations at a different time, and the file is sorted in chronological order.

Figure 11-10
Select Cases Range dialog box (time series)

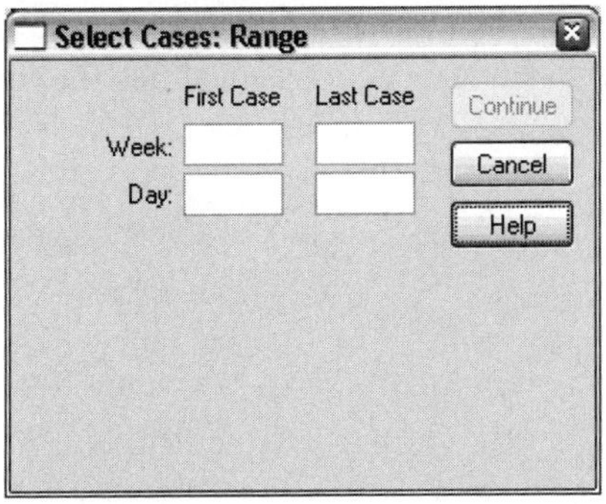

To generate date variables for time series data:

▶ From the menus choose:
Data
  Define Dates...

Treatment of Unselected Cases

You can choose one of the following alternatives for the treatment of unselected cases:

- **Filter out unselected cases.** Unselected cases are not included in the analysis but remain in the dataset. You can use the unselected cases later in the session if you turn filtering off. If you select a random sample or if you select cases based on a conditional expression, this generates a variable named *filter_$* with a value of 1 for selected cases and a value of 0 for unselected cases.

- **Copy selected cases to a new dataset.** Selected cases are copied to a new dataset, leaving the original dataset unaffected. Unselected cases are not included in the new dataset and are left in their original state in the original dataset.

- **Delete unselected cases.** Unselected cases are deleted from the dataset. Deleted cases can be recovered only by exiting from the file without saving any changes and then reopening the file. The deletion of cases is permanent if you save the changes to the data file. *Note*: If you delete unselected cases and save the file, the cases cannot be recovered.

Case Selection Status

If you have selected a subset of cases but have not discarded unselected cases, unselected cases are marked in the Data Editor with a diagonal line through the row number.

Figure 11-11
Case selection status

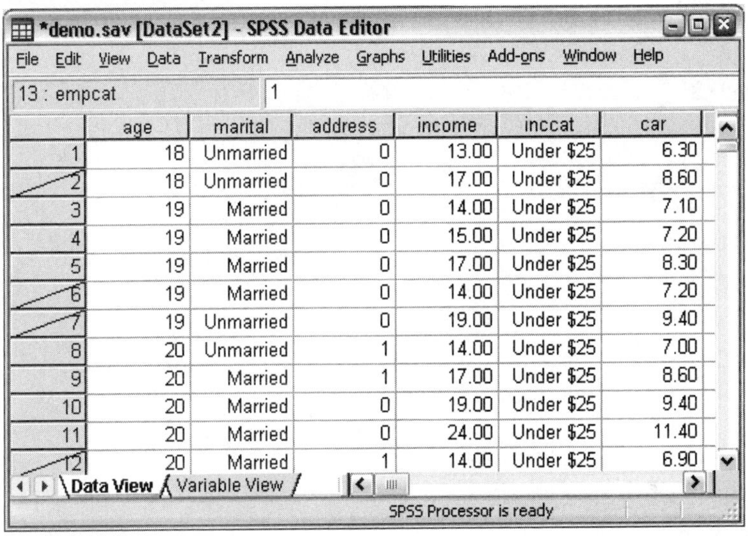
*demo.sav [DataSet2] - SPSS Data Editor
File Edit View Data Transform Analyze Graphs Utilities Add-ons Window Help
13 : empcat
1
age marital address income inccat car
1 18 Unmarried 0 13.00 Under $25 6.30
2 18 Unmarried 0 17.00 Under $25 8.60
3 19 Married 0 14.00 Under $25 7.10
4 19 Married 0 15.00 Under $25 7.20
5 19 Married 0 17.00 Under $25 8.30
6 19 Married 0 14.00 Under $25 7.20
7 19 Unmarried 0 19.00 Under $25 9.40
8 20 Unmarried 1 14.00 Under $25 7.00
9 20 Married 1 17.00 Under $25 8.60
10 20 Married 0 19.00 Under $25 9.40
11 20 Married 0 24.00 Under $25 11.40
12 20 Married 1 14.00 Under $25 6.90
Data View
Variable View
SPSS Processor is ready

12

Additional Statistical Procedures

This chapter contains brief examples for selected statistical procedures. The procedures are grouped according to the order in which they appear on the Analyze menu.

The examples are designed to illustrate sample specifications that are required to run a statistical procedure. The examples in this chapter use the data file *demo.sav*, with the following exceptions:

- The paired-samples *t* test example uses the data file *dietstudy.sav*, which is a hypothetical data file containing the results of a study of the "Stillman diet." In the examples in this chapter, you must run the procedures to see the output.

- The correlation examples use *Employee data.sav*, which contains historical data about a company's employees.

- The exponential smoothing example uses the data file *inventor.sav*, which contains inventory data that were collected over a period of 70 days.

For information about individual items in a dialog box, click Help. If you want to locate a specific statistic, such as percentiles, use the Index or Search facility in the Help system. For additional information about interpreting the results of these procedures, consult a statistics or data analysis textbook.

Summarizing Data

The Descriptive Statistics submenu on the Analyze menu provides techniques for summarizing data with statistics and charts.

Frequencies

In the chapter *Examining Summary Statistics for Individual Variables*, there is an example showing a frequency table and a bar chart. In that example, the Frequencies procedure was used to analyze the variables *Owns PDA [ownpda]* and *Owns TV [owntv]*, both of which are categorical variables having only two values. If the variable that you want to analyze is a scale (interval, ratio) variable, you can use the Frequencies procedure to generate summary statistics and a histogram. A **histogram** is a chart that shows the number of cases in each of several groups. This example will use Frequencies to analyze the variable *Years with current employer [employ]*.

To generate statistics and a histogram of the years with current employer, follow these steps:

▶ From the menus choose:
Analyze
  Descriptive Statistics
    Frequencies...

This opens the Frequencies dialog box.

Figure 12-1
Frequencies dialog box

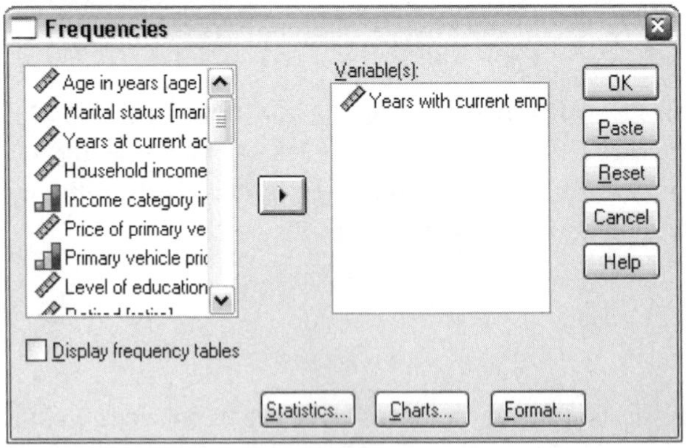

▶ Select *Years with current employer [employ]* and move it to the Variable(s) list.

▶ Deselect the Display frequency tables check box.

(If you leave this item selected and display a frequency table for current salary, the output shows an entry for every distinct value of salary, making a very long table.)

▶ Click Charts to open the Frequencies Charts dialog box.

Figure 12-2
Frequencies Charts dialog box

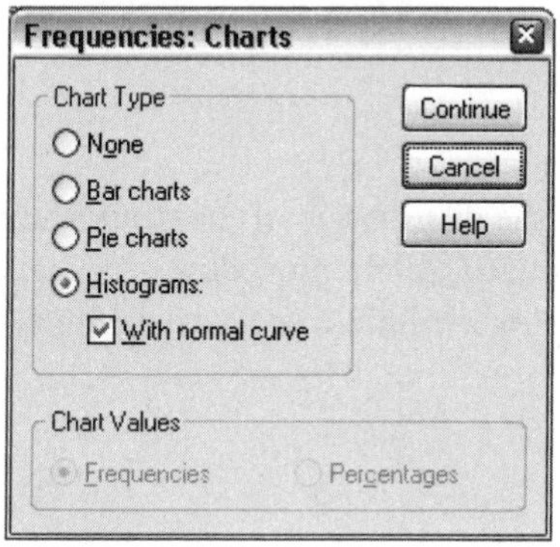

▶ Select Histograms and With normal curve, and then click Continue.

▶ To select summary statistics, click Statistics in the Frequencies dialog box to display the Frequencies Statistics dialog box.

Figure 12-3
Frequencies Statistics dialog box

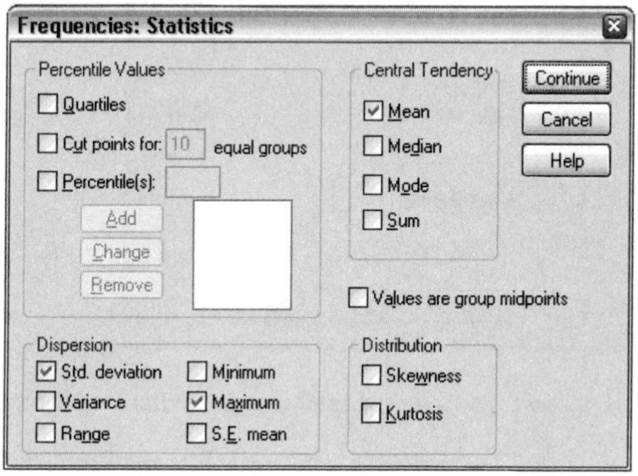

▶ Select Mean, Std. deviation, and Maximum, and then click Continue.

▶ In the Frequencies dialog box, click OK to run the procedure.

The Viewer shows the requested statistics and a histogram in standard graphics format. Each bar in the histogram represents the number of employees within a range of five years, and the year values that are displayed are the range midpoints. As requested, a normal curve is superimposed on the chart.

Explore

Suppose that you want to look further at the distribution of the years with current employer for each income category. With the Explore procedure, you can examine the distribution of the years with current employer within categories of another variable.

▶ From the menus choose:
Analyze
  Descriptive Statistics
    Explore...

This opens the Explore dialog box.

Figure 12-4
Explore dialog box

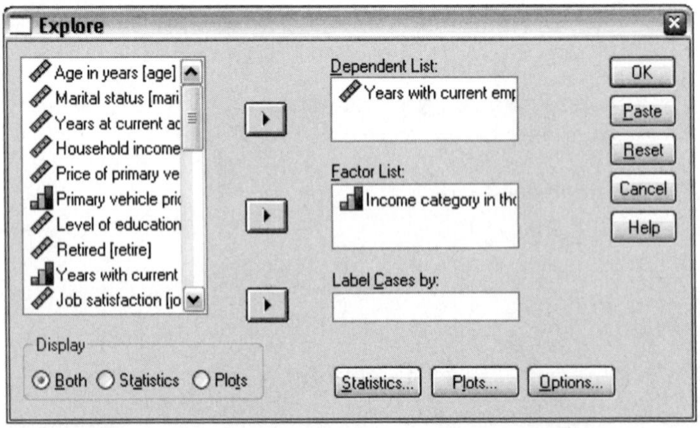

▶ Select *Years with current employer [employ]* and move it to the Dependent List.

▶ Select *Income category in thousands [inccat]* and move it to the Factor List.

► Click OK to run the Explore procedure.

In the output, descriptive statistics and a stem-and-leaf plot are displayed for the years with current employer in each income category. The Viewer also contains a boxplot (in standard graphics format) comparing the years with current employer in the income categories. For each category, the boxplot shows the median, interquartile range (25th to 75th percentile), outliers (indicated by O), and extreme values (indicated by *).

More about Summarizing Data

There are many ways to summarize data. For example, to calculate medians or percentiles, use the Frequencies procedure or the Explore procedure. Here are some additional methods:

- **Descriptives.** For income, you can calculate standard scores, sometimes called z scores. Use the Descriptives procedure and select Save standardized values as variables.

- **Crosstabs.** You can use the Crosstabs procedure to display the relationship between two or more categorical variables.

- **Summarize procedure.** You can use the Summarize procedure to write to your output window a listing of the actual values of age, gender, and income of the first 25 or 50 cases.

To run the Summarize procedure, from the menus choose:
Analyze
  Reports
    Case Summaries...

Comparing Means

The Compare Means submenu on the Analyze menu provides techniques for displaying descriptive statistics and testing whether differences are significant between two means for both independent and paired samples. You can also use the One-Way ANOVA procedure to test whether differences are significant among more than two independent means.

Means

In the *demo.sav* file, several variables are available for dividing people into groups. You can then calculate various statistics in order to compare the groups. For example, you can compute the average (mean) household income for males and females. To calculate the means, use the following steps:

▶ From the menus choose:

Analyze
  Compare Means
    Means...

This opens the Means dialog box.

Figure 12-5
Means dialog box (layer 1)

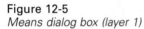

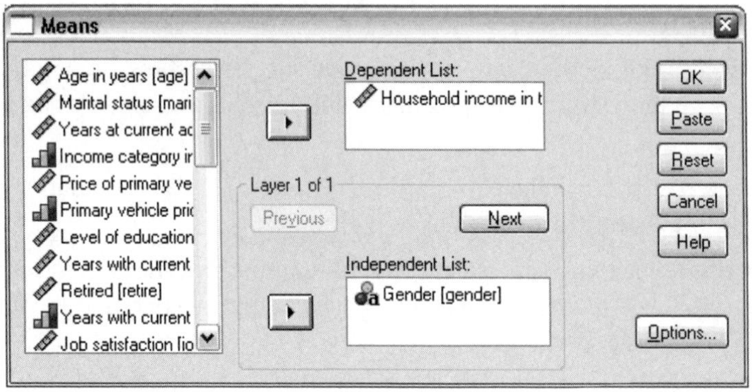

▶ Select *Household income in thousands [income]* and move it to the Dependent List.

▶ Select *Gender [gender]* and move it to the Independent List in layer 1.

▶ Click Next to create another layer.

Figure 12-6
Means dialog box (layer 2)

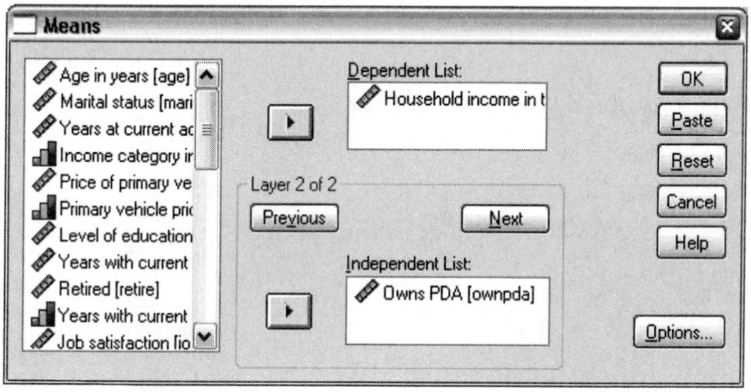

▶ Select *Owns PDA [ownpda]* and move it to the Independent List in layer 2.

▶ Click OK to run the procedure.

Paired-Samples T Test

When the data are structured in such a way that there are two observations on the same individual or observations that are matched by another variable on two individuals (twins, for example), the samples are paired. In the data file *dietstudy.sav*, the beginning and final weights are provided for each person who participated in the study. If the diet worked, we expect that the participant's weight before and after the study would be significantly different.

To carry out a *t* test of the beginning and final weights, use the following steps:

▶ Open the data file *dietstudy.sav*, which can be found in the *tutorial\sample_files* subdirectory of the directory in which you installed SPSS.

▶ From the menus choose:

Analyze
  Compare Means
    Paired-Samples T Test...

This opens the Paired-Samples T Test dialog box.

Figure 12-7
Paired-Samples T Test dialog box

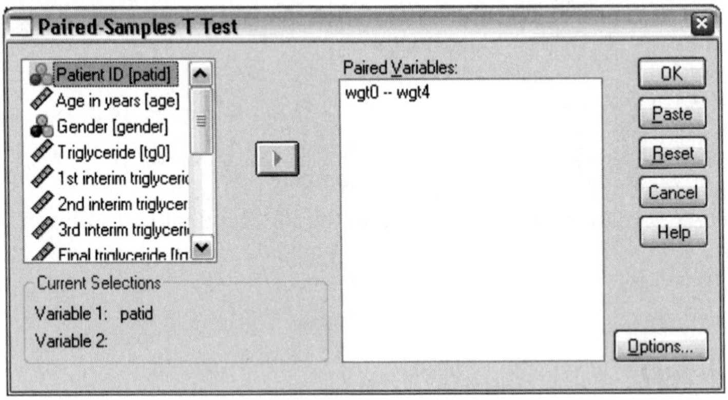

▶ Click *Weight [wgt0]*.

The variable is displayed in the Current Selections group (below the variable list).

▶ Click *Final weight [wgt4]*.

The variable is displayed in the Current Selections group.

▶ Click the arrow button to move the pair to the Paired Variables list.

▶ Click OK to run the procedure.

If there are rows of asterisks in some columns, double-click the chart and make the columns wider.

The results show that the final weight is significantly different from the beginning weight, as indicated by the small probability that is displayed in the *Sig. (2-tailed)* column of the Paired Samples Test table.

More about Comparing Means

The following examples suggest some ways in which you can use other procedures to compare means.

- **Independent-Samples T Test.** When you use a *t* test to compare means of one variable across independent groups, the samples are independent. Males and females in the *demo.sav* file can be divided into independent groups by the variable *Gender [gender]*. You can use a *t* test to determine whether the mean household incomes of males and females are the same.

- **One-Sample T Test.** You can test whether the household income of people with college degrees differs from a national or state average. Use Select Cases on the Data menu to select the cases with *Level of Education [ed]* >= 4. Then, run the One-Sample T Test procedure to compare *Household income in thousands [income]* and the test value 75.

- **One-Way ANOVA.** The variable *Level of Education [ed]* divides employees into five independent groups by level of education. You can use the One-Way ANOVA procedure to test whether *Household income in thousands [income]* means for the five groups are significantly different.

ANOVA Models

The General Linear Model submenu on the Analyze menu provides techniques for testing univariate analysis-of-variance models. (If you have only one factor, you can use the One-Way ANOVA procedure on the Compare Means submenu.)

Univariate Analysis of Variance

The GLM Univariate procedure can perform an analysis of variance for factorial designs. A simple factorial design can be used to test whether a person's household income and job satisfaction affect the number of years with the current employer.

▶ From the menus choose:

Analyze
  General Linear Model
    Univariate...

This opens the Univariate dialog box.

Figure 12-8
Univariate dialog box

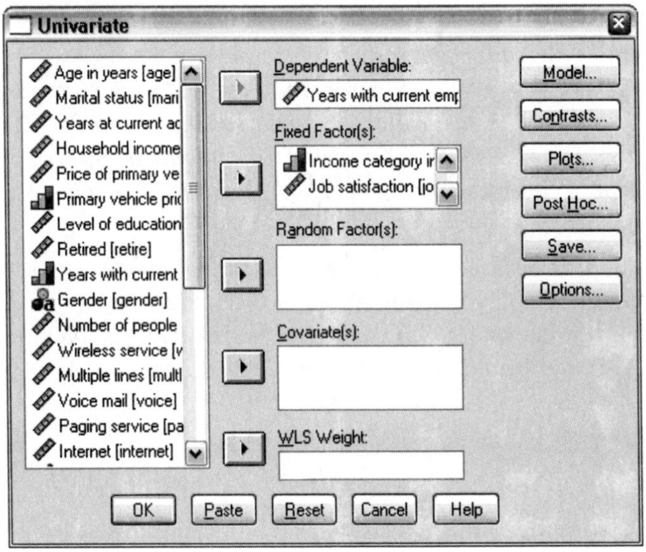

▶ Select *Years with current employer [employ]* and move it to the Dependent Variable list.

▶ Select *Income category in thousands [inccat]* and *Job satisfaction [jobsat]*, and move them to the Fixed Factor(s) list.

▶ Click OK to run the procedure.

In the Tests of Between-Subjects Effects table, you can see that the effects of income and job satisfaction are definitely significant and that the observed significance level of the interaction of income and job satisfaction is 0.000. For further interpretation, consult a statistics or data analysis textbook.

Correlating Variables

The Correlate submenu on the Analyze menu provides measures of association for two or more numeric variables.

The examples in this topic use the data file *Employee data.sav*.

Bivariate Correlations

The Bivariate Correlations procedure computes statistics such as Pearson's correlation coefficient. Correlations measure how variables or rank orders are related. Correlation coefficients range in value from –1 (a perfect negative relationship) and +1 (a perfect positive relationship). A value of 0 indicates no linear relationship.

For example, you can use Pearson's correlation coefficient to see if there is a strong linear association between *Current Salary [salary]* and *Beginning Salary [salbegin]* in the data file *Employee data.sav*.

Partial Correlations

The Partial Correlations procedure calculates partial correlation coefficients that describe the relationship between two variables while adjusting for the effects of one or more additional variables.

You can estimate the correlation between *Current Salary [salary]* and *Beginning Salary [salbegin]*, controlling for the linear effects of *Months since Hire [jobtime]* and *Previous Experience [prevexp]*. The number of control variables determines the order of the partial correlation coefficient.

To carry out this Partial Correlations procedure, use the following steps:

▶ Open the *Employee data.sav* file, which is usually in the directory where SPSS is installed.

▶ From the menus choose:

Analyze
  Correlate
    Partial...

This opens the Partial Correlations dialog box.

Figure 12-9
Partial Correlations dialog box

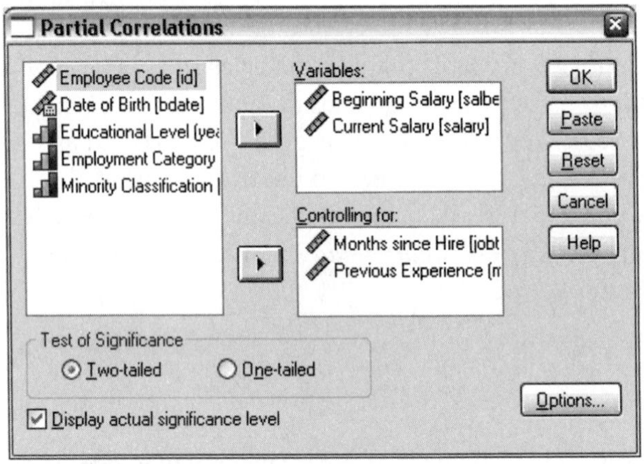

▶ Select *Current Salary [salary]* and *Beginning Salary [salbegin]* and move them to the Variables list.

▶ Select *Months since Hire [jobtime]* and *Previous Experience [prevexp]* and move them to the Controlling for list.

▶ Click OK to run the procedure.

The output shows a table of partial correlation coefficients, the degrees of freedom, and the significance level for the pair *Current Salary [salary]* and *Beginning Salary [salbegin]*.

Regression Analysis

The Regression submenu on the Analyze menu provides regression techniques.

Linear Regression

The Linear Regression procedure examines the relationship between a dependent variable and a set of independent variables. You can use the procedure to predict a person's household income (the dependent variable) based on independent variables such as age, number in household, and years with employer.

▶ From the menus choose:

Analyze
  Regression
    Linear...

This opens the Linear Regression dialog box.

Figure 12-10
Linear Regression dialog box

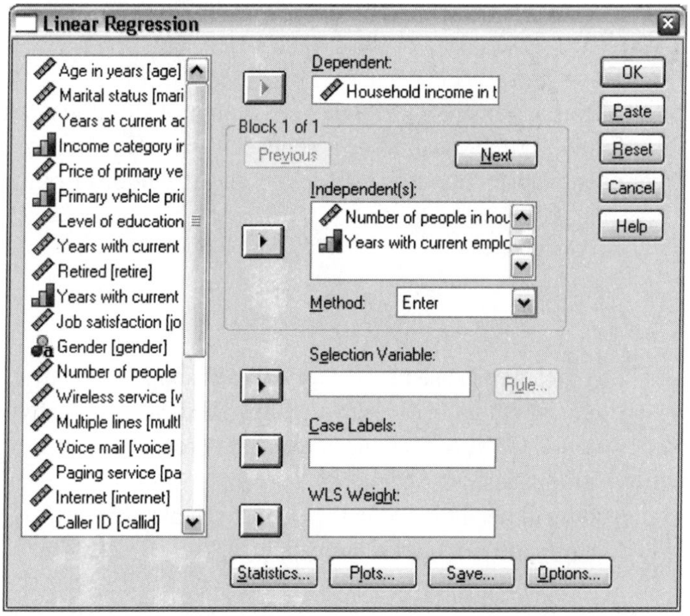

▶ Select *Household income in thousands [income]* and move it to the Dependent list.

▶ Select *Age in years [age]*, *Number of people in household [reside]*, and *Years with current employer [employ]*, and then move them to the Independent(s) list.

▶ Click OK to run the procedure.

The output contains goodness-of-fit statistics and the partial regression coefficients for the variables.

Examining Fit. To see how well the regression model fits your data, you can examine the residuals and other types of diagnostics that this procedure provides. In the Linear Regression dialog box, click Save to see a list of the new variables that you can add to your data file. If you generate any of these variables, they will not be available in a later session unless you save the data file.

Methods. If you have collected a large number of independent variables and want to build a regression model that includes only variables that are statistically related to the dependent variable, you can choose a method from the drop-down list. For example, if you select Stepwise in the above example, only variables that meet the criteria in the Linear Regression Options dialog box are entered in the equation.

Nonparametric Tests

The Nonparametric Tests submenu on the Analyze menu provides nonparametric tests for one sample or for two or more paired or independent samples. Nonparametric tests do not require assumptions about the shape of the distributions from which the data originate.

Chi-Square

The Chi-Square Test procedure is used to test hypotheses about the relative proportion of cases falling into several mutually exclusive groups. You can test the hypothesis that people who participated in the survey occur in the same proportions of gender as the general population (50% males, 50% females).

In this example, you will need to recode the string variable *Gender [gender]* into a numeric variable before you can run the procedure.

▶ From the menus choose:

Transform
  Automatic Recode...

This opens the Automatic Recode dialog box.

Figure 12-11
Automatic Recode dialog box

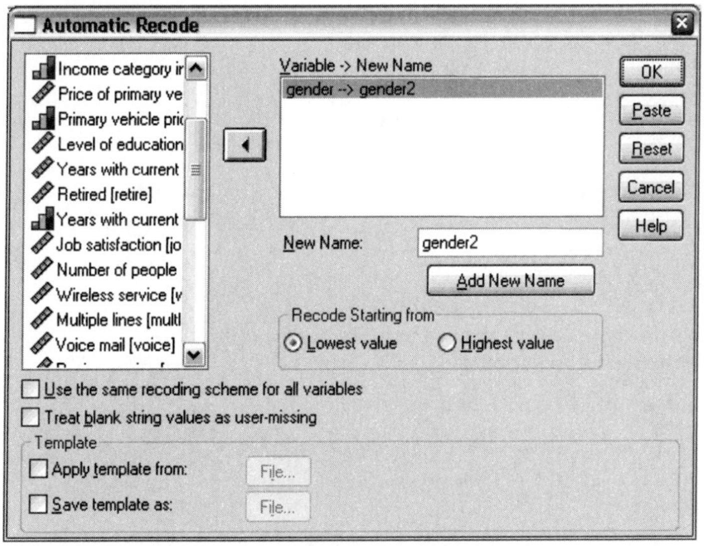

▶ Select the variable *Gender [gender]* and move it to the Variable -> New Name list.

▶ Type gender2 in the New Name text box, and then click the Add New Name button.

▶ Click OK to run the procedure.

This process creates a new numeric variable called *gender2*, which has a value of 1 for females and a value of 2 for males. Now a chi-square test can be run with a numeric variable.

▶ From the menus choose:

Analyze
  Nonparametric Tests
    Chi-Square...

This opens the Chi-Square Test dialog box.

Figure 12-12
Chi-Square Test dialog box

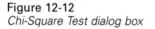

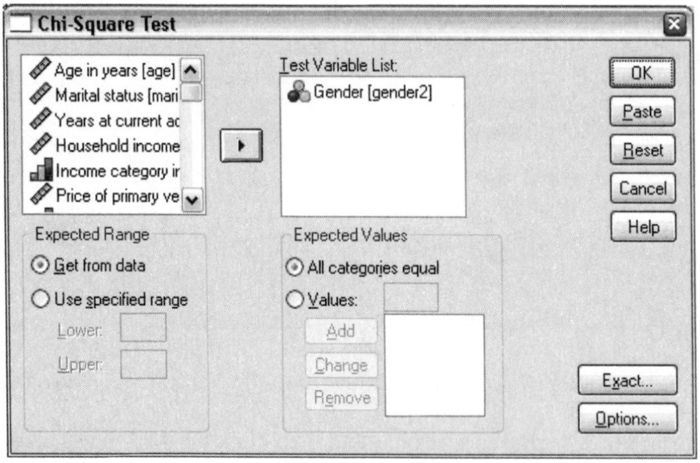

▶ Select *Gender [gender2]* and move it to the Test Variable List.

▶ Select All categories equal, because, in the general population of working age, the number of males and females is approximately equal.

▶ Click OK to run the procedure.

The output shows a table of the expected and residual values for the categories. The significance of the chi-square test is 0.6. For more information about interpretation of the statistics, consult a statistics or data analysis textbook.

Index